Earth First! and the Anti-Roads Movement

Earth First! (EF!) is one of the most controversial and well known green movements in the world and the driving force behind the anti-road campaigns of the 1990s.

Detailed accounts of major anti-road campaigns, both in the UK and internationally, are included, describing confrontations at Twyford, Newbury, Glasgow and the Autobahn in Germany, as well as information on the globalisation of Earth First!, with details of protests in Australia, Ireland, Germany, France, Holland, Eastern Europe and North America. *Earth First! and the Anti-Roads Movement* traces the origins of the movement and the history of anti-roads activism in Britain since the 1880s. EF! organisers describe how they took on their green activist identity, the launch of both EF! and the anti-roads movement in the UK, and experiences of dramatic protest. Exposing the tensions between EF! and other green activists, they explain the social and economic influences on and the culture and politics of protest.

Showing how green social and political theory can be linked to practical struggles for environmental and social change, Derek Wall investigates key topics of political and sociological interest. This is an authoritative work based on passionate and lyrical autobiographical accounts from activists, blended with a strong theoretical grounding in critical realism and social movement analysis.

Derek Wall is an Honorary Fellow at the Centre for the Study of Social and Political Movements, University of Kent at Canterbury, and a green activist.

Earth First! and the Anti-Roads Movement

Radical environmentalism and comparative social movements

Derek Wall

Routledge
Taylor & Francis Group
LONDON AND NEW YORK

First published 1999
by Routledge
2 Park Square, Milton Park, Abingdon, Oxon, OX14 4RN

Simultaneously published in the USA and Canada
by Routledge
711 Third Avenue, New York, NY 10017

Transferred to Digital Printing 2005

First issued in paperback 2013

Typeset in Perpetua by Routledge

British Library Cataloguing in Publication Data
A catalogue record for this book is available from the British Library

Library of Congress Cataloging in Publication Data
Wall, Derek.
Earth First! and the Anti-Roads Movement: radical environmentalism and comparative social movements / Derek Wall.
Includes bibliographical references and index.
1. Earth First! (Organization) 2. Environmentalism.
3. Green movement – Citizen participation. I. Title.
GE195.W34 1999 98–47372
363.7'0576–dc21 CIP

ISBN 978-0–415–19063–3 (hbk)
ISBN 978-0–415–86270–7 (pbk)

To Gillian, Larry and Vincent

There are these large pieces of metal hurtling around at high speed in residential areas. They are such a menace to life and limb that every journey made by any other means is chiefly spent dodging these monstrous objects. They are the single biggest cause of atmospheric pollution and global warming. They are the largest market for the warmongering oil industry. Their noise is the noise of the city. These 'cars' are so central to the organisation of this society, especially the organisation of work, that an illusion has to be maintained that nobody sees anything wrong with the ever increasing number of cars.

(Control 1991: 5)

Contents

Acknowledgements

This work is dedicated to Gillian, without whose support this project would have been impossible to complete, and our sons, Larry and Vincent. Gillian's energy and green commitments have inspired me in this work. Research is, perhaps, never cost-free, and this work has taken me away from my family for too many hours on too many occasions. Gillian's parents and my own also deserve a mention at this point.

Inspiration has also come, over many years, from numerous radical and revolutionary green activists. I would like to acknowledge in particular the importance of coming into contact with Larry O' Hara and (via his prison writings) Mumia Abu-Jamal, two individuals continually threatened with state violence as well as the media's gross slanders. As I write, Mumia's life hangs in the balance; with Gov. Ridge of Pennsylvania poised to sign his execution warrant any day now, we all need to keep mobilising for Mumia. Mumia is a prophet, Mama Nature's mouth piece: a lot of powerful people want to see him silenced. Protests should be mounted, and monies can be sent to MOVE care of the address at the back of this book. Larry O' Hara, another anti-racist, another green radical, another man with a gift for sharp prose, another who will not be gagged, is our Mumia. Let not a hair on either of their heads be harmed.

Like my family, my friends and comrades have been patient while I have been absent. Thanks are, of course, due to my three PhD supervisors, Professor Paul Hoggett, Peter Jowers and Dr Ian Welsh, who have supported this work, often with the kind of minute line-by-line criticism that is valuable beyond measure in documenting a piece of controversial research. All three have responded with criticism, where necessary, and the kind of creativity too rarely found in the academic arena. My examiners, Professors Andrew Dobson and Harry Rothman, have supplied equally valuable criticism. It was a good experience to have been bought a pint by Harry after the viva and to have listened to tales about his childhood and his father Benny who led the Kinder Scout Trespass in the 1930s.

David Robinson's cogent criticism, proof-reading skills and word processing know-how have been a life-saver on more than one occasion. Sarah Carty and Sarah Lloyd from Routledge deserve thanks for nurturing the project, giving support and being patient when phoned at 8.30 in the morning. Ron Price, the freelance copy-editor, and Mark Ralph, also from Routledge, worked hard to put in some finishing touches. Victor Anderson, Ian Coates, Brian Doherty, Horace Herring, Mark Reid, Paul Rogers and Chris Rootes looked at early drafts and have helped shape my thinking. The mistakes are mine – to this extent the author is not dead.

Nigel Stout and Fiona Eldridge of Mander, Portman, Woodward; Mike Kirby and Steve Cook of Ashbourne (I like the Blake artwork!); and Simon Rogers of *The Big Issue*, have all been dependable providers of nice work in times of need. Love to Gareth Thomas as well.

Finally, where would this research be without the help of Earth First!ers and others who have given up their time to be interviewed? My thanks go out to Robert Allen, Victor Anderson, Alex Begg, Martha Brett, 'Clare', Shane Collins, Chris Durham, Sheila Freeman, Davy Garland, Roger Geffen, 'Debbie Green', Steve Hunt, Daniel James, 'Jazz', Nick Jukes, Chris Laughton, 'Lorenzo', Rebecca Lush, 'Mark', George Marshall, 'Mary', 'Arthur Mix', Noel Molland, Karen Noble, Charles Secrett, John Stewart, Peter Styles, Rowan Tilly, Jason Torrance, Angie Zelter and two anonymous individuals. Alex Plows provided reflections on donga Twyford culture.

As one likes to say on these occasions, 'Long live John Africa', remembering the necessities of life beyond theoretical discourse.

Abbreviations

ALF	Animal Liberation Front
ARM	Animal Rights Militia
AU	*Action Update*
BRF	British Roads Federation
BUAV	British Union for the Abolition of Vivisection
CCTV	closed-circuit television
CJA	Criminal Justice Act
CJB	Criminal Justice Bill
CND	Campaign for Nuclear Disarmament
CPRE	Council for the Protection of Rural England
DAC	Direct Action Committee
DoE	Department of the Environment
DoT	Department of Transport
DTP	desk-top publishing
EC	European Community
EF!	Earth First!
ELF	Earth Liberation Front
FoE	Friends of the Earth
FIT	Forward Intelligence Team
GA	*Green Anarchist*
GATT	General Agreement on Tariffs and Trade
GHRA	Golden Hill Residents Association
GL	Greenpeace London
GLC	Greater London Council
HSA	Hunt Saboteurs Association
LCC	London Cycling Campaign
LETS	Local Exchange and Trading Scheme
MEP	Member of the European Parliament
NAFTA	North American Free Trade Agreement
NGO	non-governmental organisation
NIMBY	not-in-my-back-yard

NSM	new social movement
NUS	National Union of Seamen
NVDA	non-violent direct action
POS	political opportunity structure
PSBR	public sector borrowing requirement
RAG	Rainforest Action Group
RDT	relative-deprivation theory
RIA	Roads Improvement Association
RMT	resource-mobilisation theory
RSPCA	Royal Society for the Prevention of Cruelty to Animals
RSPB	Royal Society for the Preservation of Birds
RTS	Reclaim the Streets
SSSI	site of special scientific interest
SMO	social movement organisation
SWP	Socialist Workers Party
TDA	Twyford Down Association
WWF	World Wide Fund for Nature

1 Earth First!, profits last – no more roads!

Some of the younger ones began to climb on to the machine, swarming like monkeys, chanting as they climbed. This is awful, she thought. If the police arrive now it could turn nasty. But she realised she was excited by the prospect as well as anxious.

(Kingston 1996: 62)

Activists tried to close down work on Shawford and Oliver's battery sites. Met with considerable violence: punches, kicks, groping of some women, use of industrial hose....By 11:30, numbers boosted to 400, activists outnumbered workers and security guards, work ceased at Oliver's battery. No work done for the rest of day.

At 3:30...500 people marched into Winchester [city] centre shouting and singing 'EARTH FIRST! PROFITS LAST! – NO MORE ROADS'. Assembled in front of [four] Guardians of Winchester in turn – MP's office, town hall, Cathedral and College. Bags of Twyford chalk dumped on doorsteps, letters delivered. [Thirty-four] Winchester College boys shouted 'You let Twyford Down!' at college head.

(*Do or Die!* 1993 (2): 10–11)

Monkeywrenching, the Met and military intelligence?

Midday Monday, 18 July 1996, the Embankment, London, UK. I struggle down the underground station steps, sweating with the pram and my robust toddler son. We had already changed location, redirected by a familiar mobile-phone-equipped activist at Charing Cross. Arriving at a second set of gardens besides the Thames, we found an array of dreadlocked protesters, plenty of the usual suspects. Video-camera-wielding black-clad police officers filmed us. The word went round that their presence was too great for a non-violent assault on the planned target of Australia House, and we needed to regroup in a new location. While committed to the cause of drawing attention to the war in Bougainville, where Australian-backed Papua New Guinean soldiers were waging a Vietnam-style jungle war against anti-mining separatists (Quodling 1991: 1), I feared the summer: globally warmed(?) air would be too much for my 2-year-old son. I decided we should go home rather than move position again. Behind, a voice asked if I needed help to master the steps to the tube. I declined, and turned to see a uniformed police officer from the Forward Intelligence Team (FIT). Presumably because I was last to leave, I was in his eyes the likely guide to any stragglers looking for action. He asked if he could accompany us. My answer was non-committal. He followed us on the Circle and Northern (underground) Lines all the way to Camden in distant north London before realising that we were not travelling to meet Earth First!ers for the next stage of their protest: instead we had a brief shopping trip to make before going home.

FIT officers aim to identify and make contact with key activists in Earth First! and the anti-roads movement. The FIT are uniformed and open about their intentions. This officer had attempted to engage me with detailed questions about my involvement and opinions. He had been particularly keen to learn if I'd attended an earlier action where 7,000 protesters, as opposed to the thirty or forty we had been with earlier, had held a massive illegal free party on the M41, the normally busy London motorway.

Earth First!'s monthly newsletter *Action Update* (*AU*) reported the creation of the FIT in February 1996:

> The 12-strong team can now be seen on every sizeable action around the country. They aim to build a 'rapport with (key) activists so that people likely to provoke disorder can be identified early in an event'. The aim of this is to split the direct action movement from within. They are relying on people talking to them or simply grassing other activists up. They have already approached EF!ers in London asking for names of 'potential troublemakers'. At the Whatley demo they were offering sizable amounts of cash to photographers for their films.
>
> (*AU* 1996 (24): 3)

The FIT's interest in the M41 action (I'd been told by the officer that the far-right, far-left and 'football' were other areas of professional attention) is just one indication of the potential power of radical environmental action. Anti-road protests, in particular, have created an immense impact in Britain during the 1990s. Road construction has been slowed and construction bills inflated by activists variously jumping on earth-moving equipment, establishing protest camps, sitting up trees that were due to be felled, squatting buildings about to be demolished and, most dramatical of all, tunnelling beneath construction sites. A transport policy based on car use has come under sustained attack. During an eviction of the M11 campaigners in east London an activist had stated:

> [I]t was a fucking awful night, it was horizontal rain coming down, it was freezing…I happened to get up in the tree anyway because no one else was

> up there, and I was concerned that the Old Bill were going to steal a march on people and come early. I had started guarding the bottom of the ladder to make sure other people had access, and then I went up and parked myself up there.... Come three in the morning and the Old Bill come, you know, swarming in, hundreds of the bastards.
>
> ('Mix' in interview)

In 1998 'New' Labour supported a Road Traffic Reduction Act, initiated by the Green Party, the Welsh nationalist Plaid Cymru and Friends of the Earth (FoE) and aimed at setting targets for reduced car use. The media carried hundreds of news items about the protest. Anti-roads action permeated popular culture. 1998 witnessed the long-standing Granada TV series *Coronation Street*'s ersatz Manchester roads activist Spider, modelled on Swampy the Fairmile tunnel-digger, a BBC1 Jasper Carrot comedy about police infiltration of an anti-roads camp and the ITV drama *London's Burning* in which a middle-aged fireman joins a similar protest. The quotation at the start of this chapter is taken from *Laura's Way*, an item of 'romantic' fiction loosely based on the Twyford Down road camp (Kingston 1996).

Not all anti-roads activists are Earth First!ers, though Earth First! (UK) has acted as a catalyst, adapting tactics and forging new alliances to speed the growth of direct action. Formed in the spring of 1991, after several failed attempts, by two bored further-education students from Hastings in east Sussex, EF!'s parent body originated in the southern states of the USA. EF! (US) was launched in 1980, and quickly became notorious through its strident slogan 'No compromise in defence of Mother Earth' and its use of direct action, including sabotage. EF! (US) was inspired in part by Edward Abbey's novel *The Monkey Wrench Gang*, first published in 1975, which told the story of a campaign of billboard felling, bulldozer burning and dam demolition. The plot is a vehicle for a thinly disguised, no-holds-barred, sabotage handbook:

> First, they cut the fence. Then they dug out the rock ballast from beneath the crosstie nearest the bridge.... When a hole was cleared the size of an apple box, Hayduke consulted his demolition card (GTA 5–10–9), handy little item, pocket-size, sealed in plastic, which he had liberated from Special Forces during his previous career. He reviewed the formula: one kilogramme equals 2.20 pounds; we want three charges 1.25 kilograms each, let's say three pounds each charge, to be on the safe side.
>
> (Abbey 1991: 169)

The middle-aged MD Doc Sarvis, his assistant Bonnie Abzug, a dancer from the Bronx, polygamous 'Jack' Mormon Seldom Seen Smith and George Hayduke, a US military Green Beret turned alcoholic Vietcong fighter, together declare war on the American Dream. Strip-mining, cattle-grazing, roads, bridges and shopping-mall culture all feel their wrath. Riven with irony and contradiction, the novel features patriots who fight for the communists and anarchists in love with America, while the gang rides its camper van within an anti-road road narrative.

Abbey based his account on real activists who sabotaged environmentally damaging projects, turning their campaign into fiction, which was then recycled back into reality. *Ecodefense*, produced by EF! (US) co-founder Dave Foreman, was a widely distributed 'ecotage' manual, with hints on attacking the corporate offices of polluting companies, on felling and burning advertising hoardings and dismantling animal traps. Most controversial was the tactic of tree-spiking, which involved hammering nails, preferably ceramic to avoid discovery by metal detectors, deep into trees to destroy saw-mill blades and wreck chainsaws. Critics argued that malfunctioning blades skidding from an EF! (US) nail might remove a forest worker's limb. Abbey, in contrast, enthused in the Foreword to *Ecodefense* that

> No good American should ever go into the woods again without this book and, for example, a hammer and a few pounds of 60-penny nails.... You won't hurt the trees; they'll be grateful for the protection; and you may save the forest. My Aunt Emma back in West Virginia has been enjoying this pleasant exercise for years. She swears by it. It's good for the trees, it's good for the woods, it's good for the earth, and it's good for the human soul.
>
> (Abbey in Foreman and Haywood 1993: 4)

Such 'ecotage' or 'monkeywrenching' has been linked by academics and initial EF! (US) activists to a deep-ecology perspective (Dobson 1990: 64–5; Lee 1995: 18; Manes 1990: 70; Nash 1989: 191; Scarce 1990; Zakin 1993: 290), first named and formalised by the Norwegian thinker Naess. There is a large and esoteric literature relating deep ecology, variously, to the German philosopher Heidegger, to Native American beliefs and Zen Buddhism (Devall 1985; Naess 1973, 1984 and 1989; Sylvan 1985a and 1985b; Zimmerman 1994). For Abbey and Foreman 'deep ecology' simply meant prioritising 'wilderness' and valuing diverse forms of life. Such concepts were derived from a tradition of radical US environmentalism stretching back to Muir, the conservationist and explorer, in the early years of the twentieth century. Foreman is reported to have declared, with his usual flamboyance: 'John Muir said that if it ever came to a war between the races, he would side with the bears. That day has arrived' (Abbey in Foreman and Haywood 1993: 4).

Described as 'a violent and extremist wing within the deep ecologists' (Vincent 1993: 266), EF! (US) is portrayed in many accounts as an irrational group which endorsed even bomb attacks. My dismissive 1990 summary of what I then knew of EF! (US) stated:

> Earth First! espouses dramatic direct action, which includes the sabotage of vehicles posed to destroy wilderness and the hammering of huge metal pins into trees threatened by the chainsaw. Less savoury elements of Earth First! ideology include extreme neo-Malthusian praise for diseases like AIDS, for famine and for other maladies that reduce human numbers.
>
> (Kemp and Wall 1990: 11)

Some EF! (US) founders combine a love of the desert and the forest with conservative politics, over-population obsessions and simplistic naturalism. Yet it is unrealistic to view EF! (US) as simply a right-wing green group in opposition to corporations. Its image has been rubbished by a hostile media and attacked by green rivals in a surprisingly sectarian fashion. It is above all a diverse movement:

> As one who has done extensive field work and textual study of this movement, it is obvious to me that far too much attention has been given to certain statements by Edward Abbey, David Foreman and Christopher Manes...that do not represent the movement as a whole and for all time. I trace much of the 'straw-man' problem to analysis which depends exclusively on texts. Such dependence leads to preoccupation with ideas and controversies that have appeared in print, which in turn has led to serious misunderstanding of the movement....[Moreover] early but inaccurate characterizations...are perpetuated by later writers who tend to portray this movement as a group of fed-up, nature-loving, cowboy-redneck 'buckaroos'.
>
> (Taylor 1995: 27–8)

Since 1980 most EF! (US) activists have rejected right-wing politics and more controversial forms of ecotage. The late Judi Bari, a former trade union activist, joined the movement and pushed EF! (US) to the left. Foreman, the movement's co-founder, renounced much of his former conservative perspective during a debate with the veteran social ecologist and anarchist Murray Bookchin (Bookchin and Foreman 1991). Indeed, it is difficult to label Abbey or Foreman, even at their most simplistic, as right-wingers. Abbey's doctoral thesis sympathetically examined anarchism, and many of his most controversial tactics, including tree-spiking, were inspired by the activities of the turn-of-the-century International Workers of the World, aka 'the Wobblies'. Bari, who had helped set up new IWW branches to oppose logging, had much in common with her supposedly conservative opponents in EF! (US). Controversy has, though, dogged the network. By 1991, when EF! (UK) was formed, its US parent had suffered severe state pressure as well as violent attacks from opponents (Mercer 1994: 113–16; Rowell 1996: 157–67; Taylor 1995: 27). Foreman had been arrested at gunpoint during an FBI raid; other activists had been imprisoned, and Bari had been badly injured in a bomb attack. Such assaults cemented links with Native

American and African-American green radicals, such as members of the MOVE organisation (see p. 7), who had been similarly persecuted.

I first met an EF! activist in the summer of 1991. Davy Garland was an ex-Green Party and ex-Communist Party of Great Britain member, deeply attached to working-class community politics. I had already encountered Garland, in one of his previous incarnations, at a Belfast Green Party meeting and at a number of Red–Green conferences in London. Garland is notorious for his advocacy of strident militancy and had been briefly held by the police for selling the militant newsletter *Terrarist*, a pun on the Latin term for 'earth'. When representing the Green Party, I debated green politics with Garland and a member of Class War at a public meeting in the upstairs room of a Glastonbury public-house.

Garland is far from unique in his sympathy for those who defend the planet using militant tactics. Shortly after this Glastonbury encounter with Garland, I was asked to attend a mysterious rendezvous at the Star and Garter, my local smoke-filled public-house in the St Pauls area of Bristol. I met my contact, who told me to accompany him to his car. We drove around the Bristol streets, as a security precaution, in suitably film-noire fashion. As the rain beat down, he told me that his Earth First! cell had been anonymously sabotaging JCBs and other items of heavy plant in the West Country. I had 'form', having been convicted of a minor assault on a McDonald's outlet and was a long-standing Green Party radical, so I was seen to be sympathetic to his publicity purposes. At one point he opened a briefcase full of xeroxed copies of *Ecodefense*.

In 1991 I researched and wrote an article for the local magazine *Venue* and managed to interview four distinct groups of ecosaboteurs (Wall 1991b). Around the same time I met Tim Hepple. Hepple was an enthusiastic infiltrator of the neo-Nazi British National Party (Hepple 1993). He boasted to me of his life as a former military intelligence officer in Northern Ireland, and I found him engaging company. I trusted him because of his strong contacts with the anti-fascist magazine *Searchlight*, for which I had written and whose editor, Gerry Gable, I had met with to discuss the possible infiltration of the green movement by the far-right (Wall 1989). Hepple was everywhere politically. When I met him he was prominent in *Green Anarchist*, a Green Party member, an ecosaboteur and a participant in several far-right groups. Along with Paul Rogers from *Green Anarchist* Hepple invaded the stage of the autumn 1991 Green Party Conference holding a 'Planet, not Parliament' banner. He was particularly keen, he told me when I interviewed him for my article a week or so later, to meet 'other' ecosaboteurs. A Green Party member, Larry O'Hara, whom regular readers of *Searchlight* will recognise, raised the possibility that Hepple was an *agent provocateur*, whereupon my trust evaporated (see O'Hara 1993a). I would recommend that anybody interested in the resulting controversy read both O'Hara's account and the relevant back-issues of *Searchlight*. Mercer (1994) on the right and Anderson (1993) on the left provide interesting comment.

As well as meeting such colourful characters, I have participated in diverse EF! (UK) actions, including visits to road protests at Solsbury Hill and Claremont Road, street parties in Camden Town and, on the M41, quarry actions near Bristol, a demonstration against a branch of Midland Bank and a McDonald's occupation. I attended two EF! (UK) national Gatherings, in April 1995 and June 1996.

My main form of political participation during the summer of 1995 was the campaign to save the US death-row prisoner and MOVE supporter Mumia Abu-Jamal from execution. The MOVE organisation, based in Philadelphia, is a radical, mainly African-American, group which fuses religious and political themes, including green commitments (Abu-Jamal 1995: 181; Friends of MOVE 1996). MOVE is inspirational to radical greens and animal liberationists. Established by the prophetic John Africa, MOVE has a membership which lives communally, is vegetarian and committed to the fight for social and ecological justice:

> We don't believe in this reformed world system – the government, the military, industry and big business. They have historically abused, raped and bartered life for the sake of money....Industry has raped the earth of countless tons of minerals, bled billions of gallons of oil from the ground, and enslaved millions of people to manufacture cars, trucks, planes and trains that further pollute the air with their use. And because of the billions of dollars in profits to be made, the system will favour artificial transportation over the legs and feet Mama [Mother Nature] gave us to walk and run with. MOVE's work is revolution. John Africa's revolution, a revolution to stop man's system from imposing on life, to stop industry from poisoning the air, water, and soil and to put an end to the enslavement of all life.
>
> (Friends of MOVE 1996: 69)

Subject to severe persecution, MOVE supporters have spent decades in US jails, while in 1985 a MOVE household was bombed by the FBI, with thirteen fatalities including young children (Harry 1987; Walker 1988). The campaign for Mumia Abu-Jamal, an eloquent commentator on race, religion, revolution and the environment, involved organising a non-stop picket of the US Embassy, direct action against US companies and publicity work. EF! (UK) activists became increasingly active in the Abu-Jamal campaign.

My experiences of secret policemen and ecosaboteurs illustrate Earth First!'s often dramatic and sometimes sensational associations. Images of masked defenders of the planet assaulting road building machinery, thousands of trespassers partying on motorways and the sheer colour of anti-road campaigns all make 'a good story'. My interest extends beyond such descriptions of an apparently dramatic manifestation of environmental protest. By 1991 I had been active in the green movement for over a decade and had become increasingly interested in political strategy. Greens seemed to me to have a devastating critique of social and ecological ills, showing how a society based on ever-increasing economic growth is unsustainable and fundamentally unjust; they also had an attractive vision of an alternative society. Yet the movement in its very diverse manifestations seemed strategically naive, having a limited conception of how fundamental economic and political forces that thrive on continuous productivist growth and waste might be challenged. EF! (UK) fascinated me, as did MOVE and the UK Green Anarchists, because of its fundamental hostility to such forces and its ability to bypass conventional political channels. Yet more so than these groups or even EF! (US), EF! (UK), despite its near-fundamentalist commitment, has succeeded in working with very diverse groups including hedonistic dance cultures, middle-class conservationists and radical trade unionists. Functioning as, in Gramsci's term, organic intellectuals, they have helped shake the common sense of our society, promoting elements of a green ideological hegemony in Britain by, for example, fuelling doubts about our dependence on the car.

Many EF!ers are sceptical of academics' endeavours, condemning their 'dense, laboured texts', and attempts by outsiders to repackage or analyse their ideas and actions (*Do or Die* 1998 (6): 145). Phil McLeish, the last campaigner to be evicted from the Claremont Road protest against the building of the M11 (dealt with in Chapter 4) told me that academics tend to ask 'why social movements exist, [whereas] activists want to know how to win'. Political theorists in the green movement have for some time explored, among other strategic issues, the role of agency in social mobilisation (Dobson 1990; Goodin 1992). Various attempts have been made to answer the question: 'Who is best placed to bring about social change?' (Dobson 1990: 152). Greens have often conceptualised agency in terms of universal humanity threatened by the environmental dangers and other ills described by green politics. Environmental catastrophe may be seen as a global problem affecting all individuals, irrespective of social position in terms of class, gender or ethnic status, in which case it is natural to assume that the entire human species will act as the agent of green change (Dobson 1990: 152). Yet, as Dobson notes: 'It is simply untrue to say that given present conditions, it is in everybody's interest to bring about a sustainable and egalitarian society' (1990: 153). Little attention has been given by either green academics or activists to how individuals participate in green political activity and how, in the words of the prominent social movement theorist Melucci, 'a "we", a collective and hopefully dynamic movement, is formed' (1996: 382). I feel that in this contex social movement theory can be used to investigate how protest networks grow (see Roseneil 1995), so as to aid green activists to better understand how they can mobilise others to create a new society.

Examination of the emergence of EF! (UK) and the wider anti-roads movement not only sheds light on green agency formation, green hegemony and other strategic questions, but touches upon core debates in political science and sociology. Political scientists seek to understand why individuals participate in political activity; and there are those with an interest in why, specifically in the UK during the 1990s, unconventional political activity such as direct action has become more popular. Sociologists have long sought to understand how ideas and practices are reproduced and transformed over time. In the present context, my specific concern is to examine how green movements are produced and change.

Critical realism and social movement methodology

Political research is informed or 'under-laboured', explicitly or otherwise, by assumptions about what it is possible to know within the sphere of social science. While much 'new social movement' research is explicitly linked to post-structuralist or post-modernist approaches, I have used insights from critical realism to help to understand EF! (UK) and the anti-roads movement. Critical realism is an approach to knowledge that has the potential to provide a model for understanding both political strategy and developing practical research (Bhaskar 1989; Outhwaite 1987; Sayer 1984). Essentially, critical realists seek to discover generative mechanisms or processes. A critical-realist approach to strategy conceptualises power as the product of relatively enduring social processes (Isaac 1987). A successful strategy, from such a perspective, is one that fundamentally alters the nature of such processes.

Assumptions about epistemology 'lead to fundamentally different research strategies and have the possibility of producing different outcomes' (Blaikie 1993: 1). Thus it is important to make such assumptions as explicit and open to criticism as possible.

This task is difficult, for at least three reasons. First, there is no agreement at the time of writing as to the most appropriate epistemological approach to research within social science. Given the fiercely contested nature of the debate, little can be taken for granted and a researcher must justify his or her epistemological assumptions with care (Blaikie 1993: 1).

Second, establishing and justifying a coherent set of epistemological assumptions can be so onerous and complex a task as to obscure the central work of one's specific research. It would be relatively easy, if somewhat fruitless, to produce chapter after chapter critiquing social-research methodology before moving on to the drama of anti-roads action. Equally, epistemological assumptions about that which can be known must be translated into effective research methods capable of producing answers to particular and tangible questions.

Third, many researchers now question particular notions of explanation, truth, and the nature of evidence, rejecting approaches that can be termed 'realist'. This shift can be described in fairly simple terms by contrasting, at a relatively superficial level, the conceptions of (the pursuit of) knowledge exemplified by Marx and by Nietzsche. Marx argued famously that if essences and appearances coincided there would be no need for 'science', implying that social inquiry should be concerned with identifying hidden essences or mechanisms that 'explained' events (1991: 956). To Marx, it was self-evident that the study, however defined, of 'appearance' would not necessarily reveal the 'real' or 'essential' nature of a phenomenon. The knowledge of such essences, is a political process that acts as a precondition for liberation. For example, surface appearances might suggest that workers received 'a fair day's pay for a fair day's work', yet 'science' reveals that in fact the workings of the capitalist economic system means that workers' 'surplus labour power' is being misappropriated by the capitalist. Research into such exploitation might aid workers to increase their understanding of their position and so perhaps accelerate social change.

Nietzsche, in contrast, came to reject any concept of a liberating scientific pursuit of 'real' knowledge in the social sphere:

> When we examine the mirror in itself, we discover in the end nothing but things upon it. If we want to grasp the things, we finally get hold of nothing but the mirror. This, in the most general terms, is the history of knowledge.
>
> (Cited in Megill 1985: vi)

Increasingly it is argued that discourse *is* the 'mirror' through which we attempt to see social reality and that there is little or nothing outside of this 'text'. Instead of seeking to define underlying and essential social mechanisms, normally invisible to social actors, the task is to map the discourse of social actors. This is not to deny that objects exist externally to language; nor is it to say that only language is 'real'. Rather it means that everything we understand is mediated through the mirror of language: 'We are seen as cut off from "things" and confined to a confrontation with "words" alone' (Megill 1985: 2). Winch (1958) argues similarly that Wittgenstein's approach to epistemology shows that language conditions what it is possible to know, totally dominating other approaches to understanding.

Realist approaches to knowledge may be seen also as politically oppressive. In the social sciences such oppression can derive from a contradiction in the Enlightenment project identified by Kant (Megill 1985: 10). Natural scientists have sought to discover deterministic processes to explain the behaviour of the objects they study: for example, Newton argued that a series of laws of gravity could be used to predict the workings of the physical universe. Social science has sought to identify similar laws or essential mechanisms to apply to the study of human society. Yet such laws, if ever discovered, might reduce human individuals to the status of determined subjects without ethical status. Thus there is a contradiction between, on the one side, the Enlightenment conception of individual ethical beings who should enjoy rights and freedom and, on the other, a social science that predicts human behaviour in a deterministic way. Human beings, rather than being liberated by knowledge, would be imprisoned by it. Such a suspicion of science is shared by many greens, including EF! (UK) members who reject the objectification of 'nature'.

In this context, Marxism has been conceptualised as a project not of liberation but of exploitation, which borrows from Hegel to construct a deterministically teleological metanarrative. In contrast, Lyotard invokes Kant's term 'the sublime'. While this term may be seen as radically 'indeterminate' and impossible to capture, Norris (1992: 73) has suggested that it denotes that point which

> exceeds all our powers of determinate representation...where understanding comes against the limits of its power to comprehend experience – to 'bring intuitions under concepts', in Kant's phrase – that we achieve an insight into that which lies beyond the domain of phenomenal cognition, namely our existence as autonomous, free-willing agents who are not entirely subject to those laws or necessities that operate in the other (causal-determinist) realm.

There is a hint here that attempts to map discourse might themselves be considered oppressive projects that seek to capture the sublime and represent it. It has been argued that the 'sublime' is vital if individuals are to 'effect multiple connections and maximise political creativity in these dangerous times' and that the 'retreat into closure', that is, the operation and execution of a social science that captures a phenomenon or seeks to do so, will 'intensify injustice' (Jowers 1994: 200).

Thus it can be argued that the pursuit of a realist social science is practically unachievable, ethically undesirable and politically oppressive. Yet such conclusions leave a number of issues unresolved. Any study of society demands that judgements be made. Even techniques for mapping discourse, which seek to describe surfaces rather than investigate supposed essences, such as ethnographic observation or discourse analysis, demand that particular assumptions about objectivity be made. So the epistemological question of what we describe as 'real' remains:

> one can call everything 'illusion' if one wishes, just as one can call everything 'discourse' or 'text'. But this does not abolish the distinction between, say, an interpretation of the experience of being run over by a truck and the experience itself – a distinction which every language, if it is to function on something more than a purely fantastic level, must somehow accommodate.
>
> (Megill 1985: 42)

Megill's statement implies that there exist, however difficult to interpret, realities outside of language and that for language to function it must communicate something, however flawed, of such realities. Yet concepts of objective reality, and the search for generative processes, can be accused of containing a modernist conceit that sees social life as predictable and uniform. Typically, Wittgenstein simply stated: 'Causes are superstition.' In contrast I would argue that any form of social or political research demands a consideration of objectivity in the sense that some conclusions are more satisfactory or 'real' than others. If we reject the task of explanation and replace it with one of description, the thorny issue of objectivity remains: which description is the most appropriate? Cutting through this debate by asserting that there is no necessary connection between representation and the subject under examination seems to erode the possibility of undertaking 'meaningful' social research.

Norris (1992) argues that the anti-realism of 'strong' social constructivists makes difficult any principled oppositional stand on issues of local or world politics. EF! (UK) activists would argue that global environmental issues, such as potential nuclear war and the greenhouse effect, have 'real' implications for human beings and the non-human environment. How can green politics fight for the earth if practical conclusions about the effectiveness of environmental action are impossible to judge?

An additional attraction of critical realism is its attempt to investigate both 'nature' and 'society', eroding the often strict distinction between these concepts, and prioritising green political concerns. Critical realists, in particular, 'seek to put nature back into the nature–society relationship', according to Hannigan (1995: 184) who suggests that many social movement theorists ignore environmental factors to the detriment of effective analysis. Thus Martell (1994: 131) argues:

> Explanations of environmentalism can be *too* sociological. They explain environmentalism in terms of external social factors, but they too often exclude problems identified in the content of its discourse from having a bearing on the explanation of its rise.

Dunlap and Scarce (1990), examining twenty years of US polling data, argue that environmental concern varies directly with the intensity of environmental problems. While post-structuralist/constructivist accounts argue that environmental problems have to be 'constructed' by scientific research, media exposure and political reaction (Hannigan 1995: 31), critical realists argue that examination of the impact of such mechanisms should not obscure the importance of environmental processes. Indeed constructivist accounts have argued that the construction of 'problems' is likely to be more successful if 'empirical credibility' is evident and that events such as nuclear disasters have significant effects (Snow and Benford 1992: 140).

Yet to claim that the search for causal explanation and certainty in the areas of social movement or green (or other) political research is 'feasible...has a tantalizing shock value....It is a strong claim smacking of hubris: How can it be justified, both conceptually and operationally?' (Huberman and Miles 1985: 351). In this context, I would agree that there

> is neither full determination nor complete closure, but there are recurrent contingencies and causal tendencies that can be identified and mapped, so as to render some explanations more powerful, more fully saturated, than others. Finally, social behaviours and structures arise in part from motivated human acts – understandings, meanings, intentions – so that the phenomenological dimension is a necessary, although not sufficient, component of those lawful process we seek to explain.
>
> (Huberman and Miles 1985: 354)

Critical realism searches for 'recurrent contingencies and causal tendencies'; yet to suggest that reality is constituted by particular mechanisms or processes does not preclude identical events being produced by different processes (Bhaskar 1989). Equally there may be no single explanation for some or other pattern of behaviour and it may be that a series of distinct explanatory mechanisms will be relevant:

> An empirical phenomenon is explained as the result of multiple chains of causation. Explaining the effects of the 1989 California earthquake, for example, might require knowledge of a variety of different mechanisms: plate tectonics, highway and building construction techniques, traffic patterns, and American conceptions of nature, culture and civilization.
>
> (Steinmetz 1994: 203)

A central challenge for critical realism is to identify, using visible data, mechanisms which are often hidden (Blaikie 1993: 162; Outhwaite 1987: 22). It has been further argued (for example by Bhaskar 1989: 79) that a critical-realist approach to social reality recognises that, unlike natural structures, social structures:

- do not exist independently of the activities they govern
- do not exist independently of the agents' conceptions of what they are doing in their activities
- may be only relatively enduring.

Even where it is possible to identify the processes that led to the creation of green mobilisations, such processes will not necessarily be fixed. To conceive of a model or series of models that will apply universally as explanations of how individuals become involved in anti-roads action is unrealistic. Thus real processes may apply only within a relatively limited period of time and/or under particular conditions.

Critical realists believe that generative mechanisms should be studied with characteristically different tools in different knowledge areas. The appropriate tools of a chemist studying the processes that create new compounds are different from those of a political scientist investigating the processes that lead activists into green movements. Nonetheless, critical realists argue that research strategies may be similar in both natural and social sciences. Initially the researcher should, using empirical methods, identify and describe non-random patterns; theoretical work should follow such observation to construct models of potential generative mechanisms, which may be tested for their explanatory power (Blaikie 1993: 170; Outhwaite 1987: 33).

Miles and Huberman (1994) have attempted to use the insights of critical realism to develop practical techniques for examining qualitative data. Arguing that 'social phenomena exist not only in the mind but also in the objective world – and that some lawful and reasonably stable relationships are to be found among them', they suggest that 'lawfulness' comes from 'regularities and sequences that link together phenomena'. Such regularities, although not always immediately visible, they believe, can be investigated and mapped (Miles and Huberman 1994: 4). They also affirm the importance of the subjective and hermeneutical nature of human societies. It has been argued that examination of matters of environmental concern should integrate discussion of actual environmental problems with an examination of 'the processes of communication, discursive processing, normative orientation, "moral entrepreneurship" by which the antagonisms of the environmental debate get formed and transformed' (Benton and Redclift 1994: 9). Equally, Miles and Huberman argue that 'Things that are believed become real and can be inquired into', suggesting that critical realism blurs the distinction between positivist and hermeneutical approaches to social research: 'Approaches like ours, which do away with correspondence theory (direct, objective knowledge of forms) and include phenomenological meanings, are hard to situate' (1994: 5). They stress that qualitative techniques such as in-depth interviews and ethnography are vital if we are to understand social mechanisms and social meaning. Equally, they emphasise the value of an inductive approach using iterative stages, where fieldwork generates hypotheses which are developed or rejected as new stages of research are undertaken. They stress the importance of invention and creativity in research design, noting that reflexivity is important to maintain awareness of the potential weaknesses of research design. Finally, they value the use of hybrid methodologies that seek to link processes and more subjective meanings (Miles and Huberman 1994: 310).

Fielding and Fielding (1986) have adopted an approach similar to that of Miles and Huberman in seeking to overcome the strict polarity between 'objective and rigorous' research and that which is apparently 'subjective and speculative'. They argue (1986: 11):

> To produce general theory from the resolutely empirical data of anthropological ethnographers...one must look for the non-unique categories, discover or propose relationships among them, and then ask why such relationships exist.

They acknowledge the importance also of iteration and triangulation. Iteration uses a series of research stages that move towards more focused work as investigation precedes. Investigating the far-right National Front, qualitative interviews and ethnography were found to be essential in gaining sufficient 'proximity to members as to enable [their] belief-orientated decision-making to become accessible' (Fielding and Fielding 1986: 54). Yet it was felt that research could not rest with 'emic analysis', that is, a form of analysis that accepted members' own articulation of their beliefs as sufficient (Fielding and Fielding 1986: 55). By using other forms of data, such as movement texts, interviews with opponents and physical description, triangulation allowed an understanding of such controversial issues as political violence. Such a method, incorporating hermeneutics both of 'sympathy' and of 'suspicion', is challenging but also rewarding.

Theoretical perspectives can, of course, suffocate accounts of living movements with lofty, opaque and often irrelevant intellectual baggage. My approach here has been to ask movement activists how they got involved, to read movement literature and to reassess my own experience as an activist. Such research was 'under-laboured' and generally aided by a critical-realist approach to method (as outlined above) and modest use of some nuts-and-bolts social movement concepts, including networks, political opportunities, resources and repertoires.

In practical terms, I interviewed around thirty leading EF! (UK) activists, including fourteen founders of the movement plus other anti-road campaigners. Some preferred to remain anonymous; several chose to use pseudonyms; the remainder are named throughout the text. Charles Secrett, Director of FoE, and John Steward from the anti-roads network ALARM UK also granted me interviews. Victor Anderson, who organised anti-car street parties in the early 1970s, also provided fascinating critical comment. Theory was reintroduced where I felt it enhanced my own attempts to understand the phenomena. I have tried to combine historical analysis with interviews and personal recollection.

It is easy to see the anti-roads movement as a product unique to the culture, politics and environmental problems of the 1990s. In fact, far from being 'new social movements', both EF! (UK) and the wider anti-roads movement can be understood as products of long-standing green networks.

Thus Chapter 2 looks at the history of road-building, road protest and the UK green movement, with a nod to King Ludd, to provide perspectives on the action of the 1990s. Chapter 3 examines in detail the origins of EF! (US), and shows how the network came to the UK. Chapter 4 surveys the direct action anti-roads movement of the 1990s, travelling from Twyford Down in 1992 to the trials of Swampy and the other tunnel-diggers below the planned A30 in 1997. Chapter 5 focuses on activists' accounts of how they became involved, with Chapter 6 illustrating the political influences on action. Chapter 7 maps the culture and politics of EF! (UK) and the anti-roads movement. Chapter 8 shows how EF! (UK) became part of an increasingly global network of green revolutionaries. Chapter 9 outlines the lessons to be learned from the experiences of EF! (UK) and the anti-roads movement, lessons that have relevance for both green activists and academics.

2 Car wars

> It took several baton charges to clear a way through the crowd and all the time the cameras recorded faithfully the public response to the proposed motorway through the Cleene Gorge. In Ferret Lane shop windows were broken. Outside the Goat and Goblet Lord Leakham was drenched with a pail of cold water. In the Abbey Close he was concussed by a portion of broken tombstone, and when he finally reached the Four Feathers the Fire Brigade had to be called to use their hoses to disperse the crowd that besieged the hotel. By that time the Rolls-Royce was on fire and groups of drunken youths roamed the streets demonstrating their loyalty to the Handyman family by smashing street lamps.
>
> (Sharpe 1975: 38)

Introduction

Green politics and anti-road campaigns have roots. While green antecedents stretch back further, radical environmentalists in Britain in the 1880s shared many features with contemporary greens, including a do-it-yourself music culture and other arts along with an emphasis on social justice (Marsh 1982). Gould (1988) argues that an 'early green politics' was in the 1880s a significant component of what later became the Labour movement, disintegrating after 1900 but influencing British socialism until the 1940s, when Atlee's Labour governments established national parks and other modest conservation measures. During the last decades of the nineteenth century individuals such as Edward Carpenter, William Morris and Henry and Kate Salt advocated animal rights, decentralisation, local food production, sexual liberation and a range of green concerns. Typically, Carpenter, a founder of the Sheffield Socialist Society, was a Labour movement hero. The entire Labour cabinet in the 1920s signed his eightieth birthday card. Openly homosexual, he swam in Thoreau's Walden pond and translated Hindu texts into English. His socialism was that of Shelley and Blake:

> when asked how he combined his socialism with his mysticism, he answered in his gay, quaint way: 'I like to hang out my red flag from the ground floor,

> and then go up above to see how it looks' – a striking answer but not sound trade unionism. Except by people now elderly, Carpenter must be forgotten today in the labour movements he helped found.
>
> (Forster 1965)

The *Terrarist*, a one-issue publication linked to the ecotage-advocating Earth Liberation Front, carried among sabotage hints a biography of Carpenter entitled 'Gay, Green and Victorian'. Before 1900 socialism in Britain was strongly linked to green and anarchistic sentiments, and figures like Carpenter inspired thousands of working men and women. Environmental concern was manifested in other ways and among other social groups. Numerous environmental pressure groups, including the Flora and Fauna Society, the National Trust and the Royal Society for the Protection of Birds, were created between 1885 and 1900 (Lowe and Goyder 1983: 16). In turn, popular accounts of nature study sold in large numbers.

Direct action has been used by environmental activists since the nineteenth century. Indeed, the first recorded environmental pressure group, the Commons Preservation Society, created in 1865, combined disruptive direct action with legal challenges and political lobbying (Marsh 1982: 44). A special train from Euston was booked and hundreds of workers were hired by the society to dismantle railings erected around common land in Berkhamstead. Campaigners for public access to areas of countryside monopolised by aristocrats for grouse shooting used direct action with a mass trespass by Kinder Scout (Derbyshire) in 1932 (Rothman 1982; Sculthorpe 1993: 20). The occupation was controversial and its organisers, active in a Communist Party group, the British Workers Sports Federation, were imprisoned. Interviewees from EF! (UK) noted how Benny Rothman, a Kinder Scout activist, spoke at an anti-road trespass event at Twyford Down in 1993, nearly sixty-one years later. Non-violent direct action (NVDA) was used, on a larger scale, by peace campaigners in the 1950s and 1960s, influencing the tactics of greens in the 1970s and 1980s.

Some accounts of road protests may see them as part of a much broader movement of popular protest, revelry and riot (*Do or Die!* 1997 (6): 65; Searle 1997). Campaigns against land enclosure, other peasant revolts and the Luddites' destruction of technology have been seen as part of this tradition. *The Monkey Wrench Gang* which inspired EF! (US) is dedicated to the memory of 'Ned Ludd' and is prefaced with Byron's toast 'Down with all kings but King Ludd' (Abbey 1991). A lengthy article in the EF! (UK) journal *Do or Die!* (1997 (6): 65) describes the north of England's Luddite triangle in which there was a tradition of workers attacking the machinery that threatened their livelihoods (*Do or Die!* 1997 (6): 65). One commentator sees this genealogy as including even the 'Great Cheese Riot', when complaints about high prices led to whole cheeses being stolen – an act of eighteenth-century 'ethical shoplifting' similar to modern EF! (UK) actions (Searle 1997). Yet, by including such diverse popular protests within a lineage, the distinctive nature of modern activism may be obscured. Certainly, for the often vegan EF! (UK) activists, such an action as seizing cheese might be a protest more against animal exploitation than against poverty. But there can be little doubt that in order to understand the 1990s rise of direct action against the roads programme and the origins of EF! (UK) a brief excursus into the history of road building, road protest and green politics is necessary.

Prehistory

The earliest roads were trading routes, the exact origins of which are lost in prehistory. Tracks such as the Ridgeway running from East Anglia to Dorset are neolithic or earlier, dating back to at least 4,500 BC. They may have been used by early farmers transporting stone axes from Norfolk, the Cumbrian Lake District and farther afield. Complexes of enigmatic monuments including Silbury Hill and the Avebury henge, associated by some authors with Mother Earth worshipping religions, are found close to several such routes. Yet even these old roads should not be over-romanticised, for they were no doubt the same muddy tracks that transported armies as well as peasants across Britain. Road users in prehistory transformed the British landscape. There is, for example, evidence that the neolithic farmers who built long barrows and causeway enclosures near ridgeways thereby destroyed much of Britain's then extensive forests (Evans 1975).

Road development peaked in sophistication with the arrival of the Romans, 3,000 years after the end of the Neolithic Period. Charlesworth's sympathetic history of British motorways illustrates modern road-builders' fascination with the empire:

> The idea of a national road system was introduced into Britain by the Romans, who needed a network of roads, first for military purposes of conquest and the maintenance of Roman authority and, second, as Romanisation spread, for purposes of trade and general communication. It is interesting to compare a map of main roads in Roman Britain...with the motorway network of 1980.
>
> (Charlesworth 1984: 2)

In AD 60 Boudicca's Iceni tribe swept out of Norfolk, burning Roman settlements in its path and devastating Londinium. The Iceni were indulging in a tax revolt, incensed at demands to fund the erection of a temple to the Emperor Claudius, who had conquered Britain in AD 43. Yet in a sense the Iceni protest is part of a tradition of opposing road construction. While more violent than, say, protest against the M11 or Newbury Bypass, campaigns against empires have often also been campaigns against roads. Indeed, 'Operation Roadblock', a month of action against the M11 motorway was launched, with dubious historical accuracy, with the proclamation: 'The 15th of March is of course the [I]des of March when Julius Caesar got stabbed – the beginning of the end of the last great road building empire' (*Do or Die!* 1994 (4): 6). Roman road-building has acted as a symbolic resource for both motorway builders and protesters. It would be stretching the truth a little to claim Boudicca's insurrection as the first anti-road campaign in Britain, but perhaps by only a little. An EF! (UK) pirated version of a comic book entitled *Asterix and the Road Builders* took the story of Gaulish resistance to the Romans and rewrote the speech bubbles as objections to the destruction of Twyford Down by the M3 motorway (*Do or Die!* 1993 (2): 11).

If deserts follow civilisations, roads may be said to precede them. In numerous rainforest areas, whether in Queensland or the Amazon, roads bring cities and chainsaws. Typically, EF! (US) activists oppose highways as the incursion of human society into wilderness areas. Economic accumulation demands human hegemony over wilderness, with road construction as an early step. The imperial project, whether the construction of Roman Europe, Inca Peru or even Britain's first colony in Ireland, demands roads to carry troops and return plunder.

The centralised state power of Rome delivered an advanced road system, unmatched until the twentieth century. Medieval routes across the Hampshire Down land, the Dongas, supposedly named after similar tracks in Southern Africa, provided a further symbolic focus during resistance to the construction of the M3 at Twyford Down (Bryant 1996; see pp. 65–73). Sporadic national and local government action partially maintained roads until the seventeenth century; increasingly after this date tolls funded development. The typical trunk road of the eighteenth century was chaotic and muddy. First the canal network and later rail took the burden. Before the invention of motor vehicles, conservationists in the Lake District opposed the construction of a new railway from Windermere, fearing that it would bring a wave of tourists who would erode the beauty of the area. As early as 1844 the poet Wordsworth condemned the Kendal–Windermere Line, while successful campaigning prevented the railway from being extended. Ruskin, who mixed environmental concern, paternalism, prejudice and quasi-socialism, feared the coming of the tourists, stating with characteristic tact: 'I don't want to let them see Helvellyn while they are drunk' (Millword and Robinson 1970: 246). A bill proposing a railway line in Ennerdale was defeated in the House of Commons in 1884, largely through the efforts of James Bryce who feared it would damage 'the unspoilt and wild scenery' (Millword and Robinson 1970: 246). Self-interest rather than a desire to conserve fuelled the canal owners' opposition to rail routes such as the Great Western Line from Paddington, London, to Temple Meads in Bristol.

Autobahn

The first motor vehicle to use a British road was introduced just four years after the Ennerdale victory, when a three-wheeled Benz was imported in 1888 (Johnson 1974: 64). From the very earliest days opposition to the car was apparent. Until 1896 the new 'light locomotives' were so feared that they had to be proceeded by a red-flag-waving assistant as they crawled along country lanes at a maximum of 4 mph. In response, the Royal Automobile Club was created in 1897 and eventually secured an increase in the speed limit to 20 mph. In 1905 the Automobile Association broke away to pursue more militant tactics in support of motor vehicle use (Wistrich 1983: 87). Opposition to the car continued. Hamer records an illustration of a late Victorian or Edwardian anti-car campaigner displaying a poster which proclaimed:

> Men of England. Your birthright is being taken from you by Reckless Motorists. Reckless Motorists drive over and kill your children. Reckless Motorists drive over and kill both men and women. Reckless Motorists kill your dogs. Reckless Motorists kill your chickens. Motorists fill your houses with dust. Motorists spoil your clothes with dust. Motorists, with dust and stink, poison the air we breathe, thus injuring your [breathing].
>
> (Hamer 1987: 27)

Such opposition, slight as it was according to Hamer, came from those with a vested economic interest who saw the motor vehicle as a threat to the employment of livery-stable workers. Much of the modern network of trunk roads and bypasses such as London's North Circular was planned as early as 1910 and built during the 1930s (Hall 1980: 57), and pressure for increased road construction grew. By 1914 there were already 400,000 motor vehicles in use on British roads (Johnson 1974: 64).

During the 1920s new opponents of the car emerged in Britain. The deaths on the road of an annual 6,000 people forced the government to establish a Royal Commission on Transport in 1928. Such bloodshed also led to the creation in 1929 of the Pedestrians' Association for Road Safety which lobbied for the introduction of driving tests (Hamer 1987: 31).

While in the 1980s campaigns increased for bypasses to relieve traffic, during the 1930s conservationists argued for new trunk roads which would take traffic away from sleepy country lanes. Typically, the Council for the Protection of Rural England (CPRE), founded in 1926, saw trunk roads as a means of funnelling destructive traffic away from areas of natural beauty. Such an agenda was effectively identical to that of the pro-roads lobby which advocated a national network of motorways and new trunk roads. Seeking to shift the ill-effects of car use, many environmentalists, from the 1920s through to the late 1960s, legitimised the motorway. Bizarrely, the first campaigning pressure group for increased road building was launched by cyclists concerned 'about clouds of dust produced by [riding] in the days before roads were mettled'. Thus the Cyclists' Touring Club and National Cyclists' Union co-founded the Roads Improvement Association (RIA) as early as 1896 (Hamer 1987).

The British Roads Federation (BRF), founded in 1932, soon replaced the RIA. Along with ancient Rome, modern Fascist Italy and Nazi Germany provided models of motorway development (Charlesworth 1984: 11). Charlesworth's description of the Italian 'Autostrade' neatly defines the modern motorway as a road 'reserved exclusively for motor vehicles, linking important centres...being as direct as possible...to avoid built-up areas and consist[ing] of long straight sections...' (Charlesworth 1984: 11). Inspired by Mussolini's fascist economic philosophy, the 'autostrade' was part of a corporate project linking industry and state so as to accelerate national growth. Construction began in 1924 on a route from Milan, and by 1939 seven Italian motorways had been constructed.

In 1937 a 255-strong delegation, comprising 58 MPs, 54 county surveyors as well as numerous representatives of road-building organisations visited Hitler's Germany to inspect the autobahn programme (Charlesworth 1984: 14). Much German hospitality, including the Oktoberfest in Munich, was enjoyed. An address was given by Dr Todt, the Nazi minister in charge of transport, and the delegation, travelling by train, observed an historic meeting between Hitler and Mussolini. As a result the county surveyors in the delegation agreed to lobby for a motorway programme and the British Transport Minister Leslie Burgin visited Germany in 1938. The outbreak of the Second World War scuppered both the launch of Britain's autobahns and further hospitality visits.

Road construction enjoyed general support, from the left as well as the right, and was inspired in part by a green vision of decentralisation and garden cities during the 1930s and 1940s. Abercrombie, for example, who produced the plan for a series of London ringways or radial motorways, was very much part of a tradition of planning that stretched back to the visions of Kropotkin the anarchist geographer and green socialists Edward Carpenter and William Morris.

> Incorporating other equally large notions – the green belt itself, the ring of new towns around it, the belt of expanded towns beyond that and the planned migrations that would take more than a million Londoners to new homes outside – Abercrombie's vision was a grand one.
>
> (Hall 1980: 60)

Motorways, new towns and greenbelt inspired by early green visions gave rise to the landscape of post-war Britain, with traffic funnelled into major routes and blitzed urban populations moved into the countryside. Green discourse, however, was to legitimise productivist desires and the laying of concrete and tarmac.

Birth of the motorway

The 1945–50 Labour government saw a programme of motorway building, along with the formation of the National Health Service (NHS), welfare reform, generalised economic planning and major house building as elements of reconstruction necessary to renewed prosperity. Although there were disputes over nationalisation and the creation of the NHS, an emphasis on planning was broadly supported by the major political parties, businesses, the unions and the general public. Alfred Barnes, the Transport Minister, in an episode of road-building folklore, unveiled a motorway sketch map in the House of Commons tea-room (Starkie 1982: 2). The 800 mile hour-glass plan comprised routes linking the west Midlands with Bristol, Lancashire, London and the north-east. The plan, like much post-war road construction, was halted by a deep economic recession, and construction 'languished at levels barely a quarter of those reached in the late thirties' (Starkie 1982: 3). During the post-war period, cyclical increases in the public sector borrowing requirement, leading to cut-backs in government spending to manage such debt, were often more deadly to the road builders than were the waves of protest they encountered.

The mid-1950s witnessed a period of renewed pro-roads lobbying and revived economic growth. In 1955 the BRF launched a 'roads crusade' to whip up public support for motorways, and car ownership was rapidly increasing. During 1956 nine parliamentary debates on the issue, supported by the pro-roads lobby, were held. Eventually the long-delayed construction began. The very first stretch of motorway was opened around Preston in 1958 by Prime Minister Harold Macmillan. Significantly it was described as the Preston Bypass, the first of many, often billed as relieving lorry traffic, that grew into motorways or significant trunk roads, encouraging greater traffic flows and, in turn, nourishing demands for new roads. The legitimising description was hardly necessary: motorways were popular, and, 'with an election not far ahead, Mr Macmillan, of course, was not slow to convey the impression that motorways were the Conservatives' peculiar contribution to a modernised affluent Britain' (Starkie 1982: 1). The Conservative Transport Minister Ernest Marples held 64,000 shares out of a total of 80,000 in his own civil engineering firm and despite possible conflicts of interest was a keen motorways' advocate. Eventually a scandal was to result when the Marples' company was awarded the contract to build the Hammersmith Flyover (Hamer 1987). While Conservative Party links to the road-building industry remained strong into the 1990s, the Labour opposition was also keen to build more motorways, and a consensus of pro-roads corporatism reigned. In 1956, for example, the Trades Union Congress unanimously called for increased road building (Wistrich 1983: 86). Motorways remained popular into the 1960s, partly because so few had been built: the abstract promise of progress being more attractive than the concrete reality of demolished homes and bisected countryside. Equally, the car acted as an icon of post-war economic growth, personal freedom and increased prosperity. In 1960 the first lengthy stretch of motorway, the M1, linking Birmingham and London, was opened. In 1964 the election of Harold Wilson's Labour government led to a renewed emphasis on slum clearance, city planning and motorway building. Again economic downturn was to slow the programme, but motorways were rapidly built throughout the decade, and by 1972 1,000 miles of mainly rural routes linking major cities had been completed.

Pressure for urban networks of motorways in Glasgow, Leeds and London had also risen during the 1960s. In 1963 the Buchanan Report, endorsed by Marples, had proposed the segregation of motor vehicles and pedestrians in cities (Charlesworth 1984: 181). Buchanan emphasised two key ideas – hierarchical road systems and precincts for pedestrians – developed by the poetically named Alker Tripp two decades earlier. Tripp, a police commissioner, argued that 'the ultimate aim...is to make the vehicle tracks as well adapted, fenced and arranged...as the railways are' (quoted in Wistrich 1983: 66). As it had been for Abercrombie, the emphasis for both Buchanan and Tripp was on 'good environmental standards' (Hall 1980: 64). Widespread bombing of city-centres during the Second World War allowed local authorities to redesign urban areas on a grand scale, particularly during periods of intermittent economic boom. Increasingly cars and lorries were prioritised in such plans, and many buildings left standing by the bombers were flattened and replaced by tarmac and new shopping centres.

From 'not-in-my-backyard' to green protest

Between 1945 and the late 1960s anti-road campaigning was limited; yet, despite the strength of pro-road sentiments, successes were recorded. Even as the motorway programme was beginning in the 1950s, protests were mounted against the M1, forcing planners to reroute (Starkie 1982: 131). In the early 1960s the construction of the M4, which finally was to run from London to Bristol, encountered strong resistance from local community groups. Thus 'the Ramblers' Association opposed the Direct route, the Faringdon branch of the National Farmers' Union opposed the Vale route, and the Berkshire branch of the Council for the Protection of Rural England favoured the Bath road line' (Gregory 1974: 113). Typically, the Vale of White Horse Preservation Society argued that one route was too foggy, would cut through heavily populated countryside and, despite the fog, would obscure views of the Uffington White Horse. The Downs Preservation Society Committee feared an alternative route would 'interfere with race horse gallops' (Gregory 1974: 113). Motorway opponents aimed to shift the route elsewhere: campaigns that rejected the very principle of motorways, let alone the cult of the car, were at that time virtually unknown.

One important development that increased the potential for protest was the creation of the Civic Trust in 1957, principally as a result of fears that post-war urban development threatened beautiful and historic buildings (Starkie 1982: 82). Local civic societies affiliated to the Trust were a powerful force for conservation at the grassroots level, growing in number as they did from 200 in 1957 to well over 1,000 groups in the 1970s. The creation of the Victorian Society in 1958, while pre-dated by William Morris's Society for the Protection of Ancient Buildings, was another sign that conservation concerns were increasing. During the 1950s and early 1960s, these groups were rarely involved with road issues, partly because the focus of the first 1,000-mile programme was on rural rather than urban routes (Starkie 1982: 82). Yet the network of local conservation groups with 'not-in-my-back-yard' (NIMBY) motives and a largely middle-class membership was to provide recruits, often highly skilled and politically well connected, for more sustained and radical opposition to road building from the 1970s onwards.

During the late 1960s and early 1970s, opposition to motorway construction accelerated sharply and anti-car campaigns emerged. A resurgence of green politics in this period together with the construction of urban motorways that threatened thousands of home owners were the key factors that fuelled protest. During this period environmental concern grew rapidly and a number of environmental pressure groups stressing global problems emerged, including the now disbanded Conservation Society and the Friends of the Earth (FoE) (Hannigan 1995; Pearce 1991: 48; Robinson 1992: 112–14). In contrast to older conservation groups such as the CPRE, these new environmental pressure groups, including FoE and Greenpeace International, used high-profile symbolic direct action to create media attention, and so place issues on the policy agenda (Robinson 1992: 9). FoE first gained attention through its nonreturnable-bottle demonstration in May 1971, when 60 people dumped 1,500 empty bottles on the doorstep of the Cadbury–Schweppes headquarters in London (Finch and Peltz 1992: 3; Patterson 1983: 141; Pearce 1991: 50; Wilkinson and Schofield 1994: 14). Both FoE and Greenpeace, like EF! later on, were North American exports (see Chapter 8). Home-grown green groups such as the Soil Association, which advocated organic agricultural methods, also expanded at this time.

The resurgent concern for the environment of the late 1960s and early 1970s was often linked to calls for drastic socio-economic change:

> The principal defect of the industrial way of life with its ethos of expansion is that it is not sustainable. Its termination within the lifetime of someone born today is inevitable....Radical change is both necessary and inevitable because the present increases in human numbers and *per capita* consumption, by disrupting ecosystems and depleting resources, are undermining the very foundations of survival.
>
> (*The Ecologist*, *A Blueprint for Survival* 1972: 15)

Based on computer-generated data, *A Blueprint for Survival* proposed a political programme for the creation of an ecologically aware Britain and sought to create a 'movement for survival' to influence the Heath government, advocating the formation of an ecological political party that would seek to win governing power if its demands were ignored. More radical counter-cultural influences were also apparent (Shipley 1976: 204). For example, the Street Farmers sought to 'green' urban areas and actually built an 'eco-house' in London containing compost disposal, methane generators and fish ponds. The Farmers advocated a 'people's technology' that would 'couple the whole man [*sic*] and his life, involving unalienated work and community cooperation' (Dickson 1974: 132).

Kimber and Richardson (1974: 224) argued:

> It is not easy to accept that the London Oxford Street Action Committee, the Dwarves, the Street Farmers, and groups firmly embedded in the hippy world are part and parcel of a movement which includes bodies like the Civic Trust, the CPRE, and the National Trust.

Yet groups such as the Conservation Society and journals like *The Ecologist* were linking radical calls for transformation with conservative sentiments and had network ties to both the established pressure groups and the 'hippy world'. The development of the Green Party illustrates such articulations. The Party, established by two ex-Conservative Party activists from the west Midlands, was known initially as PEOPLE, before becoming the Ecology Party. By the late 1970s it was attracting former activists from the Dwarves and other counter-culturalists, including teepee-dwellers from west Wales such as Sid Rawle, who co-organised the Windsor Free Festival (Wall 1994b). Like FoE, Green Party activists have been recruited into numerous anti-road battles.

Ecological politics were dramatically well publicised during the early 1970s, with several newspapers covering the publication of *Blueprint* as front-page news. *The Times*' coverage of environmental stories increased by 281 per cent between 1965 and 1973 (Brookes and Richardson 1976; Veldman 1994). Established political figures debated global environmental issues, and such concerns were reflected in popular culture including comedies, documentaries, science fiction (Booth 1996: 3). Discussing a *Dr Who* TV series broadcast in June 1973, its producer, Letts, noted:

> *The Green Death* came about after Terrance Dicks and I had read a series of pieces in an environmental magazine, *The Ecologist*, about the pollution of the Earth by man [*sic*]. The articles were very disturbing and made me wish we could do something positive about it...we had no intention of attacking high technology or big business in themselves, but rather the attitude that the maximization of profit is the only good; that economic growth must be

> maintained at all costs.... Alternative technology, on the other hand, is to be used in the service of humanity, in the search for a more humane way of living.... *The Green Death* was a quite deliberate piece of propaganda.
>
> (Howe *et al.* 1994: 59–60)

Similarly, Kimber and Richardson (1974: 1) observed: 'the environment has insinuated itself into such programmes as *The Archers* and *Softly Softly* and has become entertainment'. According to Hall (1980: 86), the cultural climate had changed:

> The good future life of the early 1960s consisted in ceaseless mobility in search of an ever widening range of choice in jobs, education, entertainment and social life. The good future life of the early 1970s was seen in almost the reverse...in a small, place-bounded, face-to-face community.

Such a shift allowed politicised greens who opposed the car as a threat to the global environment to make common cause with local NIMBY activists. The resulting local campaigns, building on middle-class networks and utilising green political discourse linking car use to future global environmental destruction, were to win victories and, ultimately, to become a significant factor in slowing the growth of road construction during the 1970s.

Closing the motorway box

The principal battle of the new wave of protest of the early 1970s was to be fought in London. Both the construction of the M4 to Bristol, which swept away homes in west London, and proposals for a series of London ringways, based on Abercrombie's vision, led to bitter protest. The ringways plan was supported by the Conservative Greater London Council (GLC) in the 1960s and, after local election defeat in 1966, by the new Labour administration: 'This was the era when Labour believed in "Technology" to solve all Britain's problems: and the "Motorway Box" which had been handed to them on a plate was technology with a vengeance' (Rivers 1974: 150). A Chiswick action group failed to prevent the M4 being built through its area, but lessons were learnt. The Chiswick activists assisted campaigners against the widening of Falloden Road, eventually planned as part of the M1 in north London, to win a rare public-inquiry victory in 1969 (Hamer 1987; Starkie 1982). The 1966 motorway plan led to the formation of the Hampstead Motorway Action Group, established by wealthy north London professionals, including accountants and solicitors. A leading member argued that electoral influence made direct action unnecessary: the motorway was excluded from the area, and a similar scheme for south London's Blackheath replaced the proposed motorway with a tunnel:

> It was because both boroughs were rich, articulate and able – and both [included] marginal constituencies...with our 1,300 members and thousands of supporters, we held the key to the constituency. Our MP had a majority of only 442 and we had the power to decide which party would win.
>
> (Rivers 1974: 156)

Less well-resourced campaigners took a less constitutional approach. The official opening of the Western Avenue portion of the M4 in July 1970 was accompanied by vigorous disruption:

> Shouts of 'Philistine!' and 'Get us rehoused now!' met the arrival of Michael Heseltine, Parliamentary Secretary, at the Transport Ministry. Arriving by lorry the wrong way up an 'unopened' slip road and evading a police block, protesters from Walmer Road and Pamber Street advanced down the motorway causing total confusion among the procession of official cars. Some sang 'uncomplimentary songs, especially composed for the occasion', according to the *Kensington Post*.
>
> (Duncan 1992: 16)

City-wide concern was such that an anti-roads political party, 'Homes Before Roads', formed by the London Motorway Action Group (LMAG), fielded seventy-three candidates in the Greater London Council election, gaining 71,121 votes (Wistrich 1983). By 1971 the LMAG, was made up of 4 civic societies, 8 residents associations, 13 action groups, 2 rate payers associations and a property owners association. Predominantly local middle-class amenity groups had created a dynamic protest movement in the capital. 'Homes Before Roads' was one factor that helped to transform the initial enthusiastic support of the GLC to a position of hostility, following fierce internal debate. The ringways plan was rejected on the first day the Labour GLC sat after being elected in April 1973. Plans to resurrect elements of the plan emerged in the late 1980s and were again defeated: 'the most important point [about the London motorway network] is that it does not exist...some forty miles, out of 350 once planned. These fragments terminate arbitrarily at junctions that lead nowhere' (Hall 1980: 56). The London protest was one of a number of successful actions against urban road building in 'Cardiff, Bristol, Carlisle, Reading...Nottingham and Southampton' (Lowe and Goyder 1983: 98). Hall, himself a ringways advocate, admitted: 'Some, who might once have supported the idea of motorway planning at an abstract strategic level, suddenly found that the result threatened their home or their neighbourhood' (1980: 83). Direct action was unnecessary in these latter struggles, as the weight of public opinion surprised local politicians and crushed their plans.

The early 1970s also saw a number of anti-car actions by green activists, concerned not only with conserving a local area but with limiting the pollution and social dislocation created by motor vehicles in general. The 3 December issue of *Peace News* (1971) reported that 440 cyclists had taken part in a 'bike-in', organised by Cambridge Conservationist Society, which blocked streets. The second week of February 1972 saw the start of a month of action by Manchester Non-Violent Action Group to promote free public transport, while the counter-culture group the Dwarves (inspired by Dutch eco-anarchists), along with the Young Liberals, held an anti-car march along the route of the M12 in Redbridge (Anderson in interview).

An apparently innovative anti-road action of the 1990s has been the street party, where EF! (UK) and other activists close a stretch of busy road and hold festivities. Yet such actions, far from being new, were initiated in the early 1970s. In 1971 Victor Anderson, who now works as a researcher for Plaid Cymru MP Cynog Dafis, organised a Reclaim the Streets action that attempted to block London's Oxford Street. A second action in the spring of 1973 was held in Piccadilly Circus. These events were launched by 'Commitment', a Young Liberals group, but drew upon the experience of the peace movement's NVDA in the 1950s and 1960s. *Peace News* carried details of the action on 17 December 1971, proclaiming the event to be 'the greatest street party London had ever seen'; its aim was to demand free public transport and car-free streets, and would act as a 'symbolic action against the car's domination of the city centre...a publicity stunt'.

Anderson recalls:

> These actions drew upon emerging networks of green activists and counter-culturalists; we had links with people like the Dwarves, [and] a magazine called *Street Farmer*. I mean, we didn't have a lot to do with them but that was all part of what was going on, the *Peace News* thing...and then there were local independent environmental action groups who organised part of the demonstration....There was PEST – Planet Earth Survival Team – somewhere in north London. There were people responding to the issues, setting things up on their own. This is before Friends of the Earth.
>
> (Interview)

The idea of blocking roads was inspired by the tactics of the peace movement of the 1950s and 1960s, particularly those of the radical Committee of One Hundred. Anderson commented: 'I think the Committee was an influence in some of the things we wrote. We said that there should be a Committee of One Hundred for the environment.'

The Oxford Street action was deemed a partial success by Anderson, as eventually cars were restricted and access allowed only for cyclists, public transport and taxis. Such actions, though, were tiny in comparison with the Reclaim the Streets street parties of the 1990s at which approximately 45 arrests were made from around 200 participants. Such participants were 'people in their

20s and 30s, in white-collar professional-type jobs, [and] some students'. The action was inspired by the anti-motorway campaigns of the period:

> It was quite clear then that car was the main problem and that streets were getting more congested....In central London there was already more pollution, and there was argument in London about the ringway system and the 'Homes Before Roads' campaign, so it was already clearly a problem then.
>
> (Anderson in interview)

Reclaim the Streets and similar actions ceased after 1973, although 'bike-ins' continued in a sporadic fashion up until the emergence of EF! (UK) in 1991.

The prophet

Disruptive direct action in the form of the immobilising of vehicles was used briefly in the anti-juggernaut campaign of the early 1970s (Lowe and Goyder 1983: 62). Again the Young Liberals, including Anderson, were involved, and attempts were made to prevent heavy lorries entering Britain. Expeditions to sabotage vehicles involved smashing their windows in the dead of night terrified Anderson, who feared that parked lorries would contain angry drivers. Since the late 1960s more conventional campaigning had been supported by the *Sunday Times*, and in 1973 the backbench Conservative MP Hugh Dykes' Private Member's Bill was passed allowing lorry bans in some areas (Wistrich 1983: 124).

While these dramatic acts of direct action were on the wane, protest against motorway building accelerated during the 1970s and had a major impact on transport policy. Transport 2000, a pro-public transport group, was founded by the Railway Industry Association, the rail unions and environmentalists (Lowe and Goyder 1983: 34). In 1973 a number of environmental pressure groups, including the Civic Trust and the CPRE, created the Transport Reform Group to halt the motorway programme and demand a change in national transport policy, and in 1975 FoE launched a transport campaign promoting cycling and opposing the car.

The Midlands Motorway Action Committee, an off-shoot of the Transport Reform Group, fought the M42 plans, arguing that the motorway was unnecessary and would be environmentally damaging. The M42 public inquiry refused to consider questions of need, and the inquiry's inspector refused protesters the right to cross-examine Department of Transport (DoT) witnesses on traffic forecasts (Starkie 1982: 133). Instead of arguing against a particular route, disparate local protesters for the first time united to oppose the very notion of a motorway. At the same time there was a growing suspicion of public inquiries.

John Tyme, a Sheffield Polytechnic lecturer in planning, took a leading role in formulating tactics of direct action against inquiries. Tyme was the Conservation Society's west Midlands transport campaigner. The Society, founded in 1966, stressed wildlife loss and resource shortages, and was almost obsessively concerned with population growth. Although ideologically on the right, it placed environmental problems in a political context and perhaps for the first time stressed the increasingly global nature of ecological concern. The society acted as a particularly significant bridge between middle-class conservation groups such as the civic societies and more radical green campaigners like the Dwarves. Tyme, like the M42 activists, argued that public inquiries should not merely discuss alternative routes but decide whether traffic forecasts and cost–benefit considerations justified the demand for new routes at all (Hamer 1987; Tyme 1978). His supporters physically disrupted a number of inquiries which refused to consider such questions of need and wider environmental impact. Typically, Tyme would try to speak, the inquiry inspector would refuse to hear his objection, and Tyme's supporters would then drown out the inspector with slow hand-claps. Finally police and private security guards would attempt to eject protesters. The result was generally chaos. An Aire trunk-road inquiry in Shipley town-hall, Yorkshire, was held behind locked doors to overcome Tyme's opposition, but protesters burst through the doors.

> The chairs splintered, the handles snapped, and the first person to come flying into the chamber was Mr John Burnhope, pig farmer. Almost immediately the room filled with objectors who began singing, 'We shall not be moved'.
>
> (Tyme 1978: 28)

By linking radical environmental concerns to local campaign objectives, opposition was mobilised; and by combining legal challenges with disruptive NVDA, several inquiries were adjourned indefinitely (Dudley and Richardson 1996; Hamer 1987: 68; Starkie 1982: 133). The use of such an approach on the part of an activist within the Conservation Society, the oldest, rather staid and most conservative of the 'new' environmental pressure groups, also illustrates considerable frustration with the public inquiry system on the part of middle-class activists perceived by many policy analysts as having been successfully 'integrated' and placated by such a process (Rüdig and Lowe 1986).

Tyme was successful, in part, because he could link the language of local campaigners like the civic societies with the global environmental concerns increasingly voiced during the 1970s by more politicised greens. In contrast to earlier campaigners, he delivered a message of anti-motor car fundamentalism:

> None of our national enemies have so mutilated our cities, undermined the long-term economic movement of people and goods, destroyed our industrial base, diminished our ability to plan our community life and reduced our capacity to feed ourselves.
>
> (Tyme 1978: 1)

Tyme argued in particular that ever-increasing road traffic would lead to the depletion of precious natural resources. Such a perspective gained support with the 1973 oil crisis, when Middle Eastern producers limited petroleum supplies, forcing prices upwards and creating economic crisis in Britain and other Western economies. Tyme also emphasised what he perceived to be a democratic deficit, captured in the title of his book *Motorways Versus Democracy*, arguing that the public inquiry system was unconstitutional if it failed to allow consideration of all aspects of a road project. Tyme moved from one public inquiry to the next during the mid-1970s, with a set repertoire that used direct action to close such events when they failed to consider the desirability of the motorways as such. His approach had considerable impact. The Archway Road inquiry in north London was successfully disrupted on several occasions and the plan to widen the route into a motorway was abandoned. Quoting the *Hampstead and Highgate Times*, Wistrich noted:

> Objectors who had been enlivening proceedings in the main hall with puppet shows, sale of food in aid of their legal fund and angry shouts, struck up on a piano and a Renaissance rebec. Objectors danced showering a confetti of torn-up paper.
>
> (Wistrich 1983: 14)

The inspector retreated to a secure room known as 'the bunker', as protesters attempted to attend the inquiry, and councillor Ken Livingstone alleged that he had been punched by a DoT official (Wistrich 1983: 15). Three lines of police protected the inquiry, but the inspector called a four-week adjournment, and was never to return.

Discussing the inquiry into the construction of the M3 at Winchester, which was eventually to cut through Twyford Down, Sieghart noted:

> When not only professional campaigners, but the headmaster of an ancient public school and a university professor resort to civil disobedience to disrupt a public inquiry in order to prevent it from being held at all, something has evidently gone very wrong.
>
> (1979: 2–3)

In a similar vein Wistrich argued:

> The participants, often to their own surprise, found a militancy far removed from their quiet professional lives and hitherto associated by them with the wilder kind of political activists: these middle-class residents were, in short, unknowingly radicalised.
>
> (Wistrich 1983: 15)

Thus when Twyford Down Association's David Croker, a former Conservative Party councillor, contacted EF! (UK) in 1991 (information given in Torrance interview), he was, despite this apparently disloyal act, following a tradition of anti-roads direct action (Lamb 1996; Sieghart 1979: 2–3). In 1976 Croker had helped Tyme disrupt the Winchester public inquiry into the construction of the M3 that was to eventually cross the Down (Tyme 1978). Tyme's efforts were satirised in Tom Sharpe's comic novel *Blott on the Landscape* (1975), which tells the story of a road protest in the rural Midlands. The servant Blott is a European equivalent of the Monkey Wrench Gang members in Abbey's novel. Unlike Tyme's, Blott's efforts are not limited to disrupting the inquiry: using direct action, albeit of a violent and bizarre form, he arms himself and pours oil on the soldiers trying to oust him from his fortified tower.

The motorway programme of the 1970s was slowed considerably by anti-road activists. Public support for motorway construction became more muted, while the DoT claimed that the 1970s protests had contributed to a 'reappraisal of the basic justification for the roads programme' (Lowe and Goyder 1983: 60–1). Legal changes that allowed public inquiries to use government White Papers to justify the road-building brief cut away Tyme's constitutional arguments as a basis for protest (Dudley and Richardson 1996). Protest, of course, was not the only reason for the reduction in road construction. Economic problems, including savage public-spending cuts after the International Monetary Fund had stepped in to aid the UK economy in 1976, had a major influence. High petrol prices, despite the exploitation of (British) North Sea oil in the late 1970s, had an impact, too. By 1978, road protests had been replaced as a focus for direct action by anti-nuclear power campaigning. Tyme retired to Stroud, and new legislation to ease the progress of public inquiries was introduced.

The 1980s

Increasingly during the early 1980s the peace movement absorbed the direct action energies of green activists worried by Cruise and Trident missiles (Parkin 1989: 222; Taylor and Young 1987: 294), anti-road campaigning declined further. While the main peace pressure group, CND, dominated by the traditional left, showed limited sympathy for green politics, increasingly peace movement radicals saw themselves as 'greens' (Jones 1986). Although divisions occurred, acts of NVDA including the creation of peace camps became widespread (Taylor and Young 1987: 294). These were promoted strongly by feminists and greens who attempted to introduce new critiques to the peace movement by linking the threat of nuclear weapons to nuclear power, patriarchy, technocracy, centralisation and economic growth (Jones 1986; Roseneil 1995). The most prominent product of this critique was the women's peace camp at Greenham Common, the eventual site of Cruise missiles. Such camps were imitated by other movements of protest, with, for example, coal-miners' wives using them during the strike of the mid-1980s and animal liberationists establishing an occupation outside Porton Down's military laboratory in Wiltshire.

Green anarchism, prominent in the 1880s and resurgent in the 1970s, enjoyed modest growth during the early 1980s. In 1983 anarchists in the Greenpeace London network, not to be confused with Greenpeace International, co-ordinated the 'Stop the City' event at which 1,500 activists tried to disrupt London's financial centre (*Green Anarchist* 1984 (1): 14–15). In 1984 the *Green Anarchist* (*GA*) newspaper and network were created; *GA* was to become an early publicist for EF! (US) as well as being active in the anti-roads movement of the 1990s. For *GA* direct action was to be directly disruptive and prefigurative of 'a self-managed society' rather than simply a lobbying tactic (*Green Anarchist* 1984 (1): 1). In contrast to other greens and environmentalists, *GA* has often criticised NVDA as passive and symbolic. An account of a 'Stop the City' action celebrated the fact that 'altogether 70-odd people were arrested over the day...calling cards (bricks, paint, glue) were left at a number of financial and other relevant institutions'(*Green Anarchist* 1984 (2): 2).

In contrast to the 1970s, when there was far less communication between animal liberation campaigns, environmentalists, peace activists and others, a self-described 'green movement', albeit fractured and uncertain, began to emerge. The 'movement', though, remained weak. The majority of radical political activists remained in the Labour Party and, despite the growing radicalism of the green movement, the left perceived green concerns – with the exception of nuclear disarmament and, to a lesser extent, opposition to nuclear power and concern about the effects of transport – as somewhat trivial and 'conservative' in nature (Rootes 1992; Rüdig and Lowe 1986). On the political right, the Falklands' conflict apparently strengthened Mrs Thatcher's declining popularity, helping her to win the 1983 General Election (Parkin 1989). Those in the Party with environmental concerns, such as MEP Stanley Johnson who had written one of the earliest accounts (Johnson 1981) of environmental politics in the 1970s, were perceived as 'wets', a Thatcherite term for centrist opponents, and marginalised. The green movement seemed irrelevant to the major political conflicts of the early and mid-1980s. The peace and animal liberation campaigns, though vibrant, failed to influence the policy process; the environmental pressure groups were marginalised and green electoral results were poor (Parkin 1989: 224). Thus increasing ideological coherence and strong networking, leading to notions of a collective identity for the green movement, did not equal 'success'. Transport was a partial exception, with the left-wing Labour GLC and other radical local authorities such as south Yorkshire's, inspired by Transport 2000, boosting public transport and limiting night-time lorry traffic in London during the early 1980s.

Road construction slowly increased through the decade. A number of battles were fought, with victories on either side. FoE's Joe Weston defended Otmoor –

an area of habitat-rich marshland near Oxford – from incursion by the M40 by dividing the land into tiny packages that were to be sold to protesters so as to make compulsory purchase difficult (Lamb 1996: 8). Large road schemes cut through National Parks at Okehampton in Devon and Gargrave in north Yorkshire (Whitelegg 1989: 205). In 1980 the Armitage Report, *Lorries, People and the Environment*, called for an increase to 44 tonnes in the permitted weight of lorries. In 1982, after parliamentary protest and public disquiet, the legal weight was set at 38 tonnes (Wistrich 1983: 124).

The thrust of Conservative policy was to champion the motor car as a spur to growth and personal freedom, while privatising public transport. By the mid-1980s the legacy of Tyme had been forgotten and the DoT was proclaiming its enthusiasm for the new routes. A 1987 DoT document triumphantly stated:

> The M25 is complete. It has been one of the most ambitious and significant engineering projects undertaken in this country ... at a cost of nearly £1000 million. In total 108 motorway and trunk road schemes in England have been completed, an investment of nearly £1.5 billion [since 1973].
>
> (Quoted in Whitelegg 1989: 204)

There were new proposals for inner-London motorways, with assessment studies outlining a plan put together by the DoT published in 1988. The assessments looked like a restoration of the ringway plans of the 1970s (Stewart *et al.* 1995), and threatened tens of thousands of houses. These preliminary plans enraged hundreds of local community groups, and Transport 2000 called a co-ordination meeting. This lead to the creation of All London Against the Road Menace (ALARM), with 150 local groups involved, including cycling campaigners, FoE branches and local amenity bodies. Protest stunts such as the delivery of a Valentine's cake to Cecil Parkinson, the government minister with responsibility for the assessments, helped destroy the credibility of the proposals, which were withdrawn in May 1990. ALARM helped to defeat the proposals at the planning stage; the public inquiry stage, at which Tyme had favoured action, was never reached. According to Roger Geffen, who was active in both ALARM and the later EF! (UK): 'They dropped the lot. They also endorsed the 1,000-mile cycle route: we thought we were winning' (Geffen in interview).

Despite this victory, calls for renewed road building continued. The 1989 *Roads for Prosperity* White Paper projected a doubling of the rate of trunk-road and motorway construction, peaking in 1992. Its supporters argued that this would be the greatest programme of road building since the Romans. Parkinson again utilised green discourse, suggesting that bypasses would conserve villages and promising that trees would be planted along the new motorways.

The rise and fall of green politics

An early portent of the 1990s anti-road direct actions had been the Stroud tree campaign in 1989. A number of oak trees were to be felled to make way for an approach road to a new Tesco supermarket in the Gloucestershire town. To prevent the oaks being cut down, activists climbed them. The tree-sitters were victorious and the new Tesco had to make do with a road that weaved around the trees. Success was a product of a number of factors. Stroud was almost unique in being a local government Green Party stronghold: Party members sat on the district council and controlled the town council. The Green Party took part in the protest, thus raising its profile. Tactics were borrowed from the local peace movement. Finally, environmental concern reached record levels in 1989 making difficult the felling of trees in the locality.

Media interest, green movement membership, polling data and voting figures all indicate a dramatic increase to the status of environmental concern in the late 1980s (McCormick 1991: 107; Robinson 1992). Green Party membership rose to 20,000. Friends of the Earth grew from 31,000 to 125,000 members and Greenpeace from 150,000 to 281,000 (Frankland 1990: 13). Prime Minister Margaret Thatcher, in a sharp break with earlier statements, endorsed environmental concerns in a number of high-profile speeches in 1988, giving an 'unprecedented respectability to their articulation' (Cracknell 1993; Rootes 1995: 70). Charles Secrett, Director of FoE in the mid-1990s, noted that

> following on from the great orgasm of green concern that exploded at the end of the 1980s there was a public, political and media climate of opinion...that said we cannot go on as we are. The development path that humanity has set for itself is leading us to...environmental, and related social, economic and political, crises and we have to find a new way.
>
> (Secrett in interview)

In the 1989 European parliamentary elections the Green Party won 14.9 per cent of the vote, a ten-fold increase on its 1987 (UK) General Election figure (Rootes 1995). At the international level, the UN promoted the concept of sustainable development and the European Community, drawing upon such discourse, insisted on stronger environmental regulation. Mrs Thatcher sacked Minister of the Environment Nicholas Ridley, who had been notoriously hostile to concern for the environment. The policy-making process was perceived to be open to environmental demands, and green consumerism was being widely advocated (Doherty and Rawcliffe 1995: 241; McCormick 1991: 107; Robinson 1992: 191). Although this 'new green society' saw global warming, rainforest destruction and ozone depletion as problems, the need for wider political and cultural analysis of the causes of such problems was largely unrecognised (McCormick 1991: 107–27). Green political demands were being choked under the weight of Thatcherite assumptions:

> The rise of 'green' consumerism, the acceptability of 'green' capitalism, has to some extent been a compromised response to Thatcherite policy and philosophy. It produced a 'greenness' which suited both the markets and the Conservative Party. Private sponsorship of environmental groups, ethical investment and even the slick marketing of the Green Party 1987 election manifesto are all signs of the strength of Thatcherism.
>
> (Robinson 1992: 191)

Typically arguing that it had a major impact on the policy process, FoE further claimed that 'environmental concern [was] professed by government, commerce, media and public in the late 1980s' (Huey 1990: 6). 'In short', claimed Weston, 'Friends of the Earth has moved from being the amateur, evangelical, fundamentalist ecocentric pressure group of the 1970s to a professional pragmatist organization which is run virtually like any other modern company' (1989: 195). The energies of those who sought to create a green movement had been largely incorporated in conventional forms of political and social participation (Doherty and Rawcliffe 1995: 242; Wilkinson and Schofield 1994: 133). By 1990 the Green Party was suffering declining membership: there had been a massive fall in poll ratings, and severe factional conflict was endemic. UK environmental policy remained substantially unchanged, public interest fell and pressure groups lost support (Rüdig 1993; Wilkinson and Schofield 1994):

> [A]s quickly as it had risen in 1988, the environment dropped down the political agenda in the run up to the 1992 General Election. During 1991, as other issues, including the economic recession, the poll tax and the Gulf War, began to force the environment...off [both] the front pages and the political agenda, a number of newspapers began to lay off their specialist environmental correspondents and coverage of environmental issues in other types of media fell. In turn, mobilisation of members and resources among environmental groups slowed and, for some, income began to decline.
>
> (Doherty and Rawcliffe 1995: 241)

After several years of green movement growth, radicals were normalised, realists marginalised and stagnation had set in. In short, by 1991, when EF! (UK) mobilised a revival of anti-road actions in response to *Roads For Prosperity*, the recreation of green radicalism had become a necessity for the green movement family.

An old social movement

Much new social-movement (NSM) analysis is hostile to claims to historical continuity, while greens often proclaim contemporary environmental movements 'new' (Wall 1994a). Within the social sciences, Marx's emphasis on historical process has increasingly been replaced by a sympathy for Foucault's insistence on discontinuities and Lyotard's rejection of 'metanarratives' (Megill 1985). The theorisation of NSMs can be linked to the apparent demise of Marxism and to the need to seek out new historical agents beyond the organised working-class (Scott 1990). Typically, Laclau and Mouffe (1985) have sought to construct a politics based upon the articulation of 'new' movements. The 'largely cultural character of new social movements, their loose organizational structure, and their emphasis upon life-style, rather than conventionally political issues', are all features of the anti-road protests of the 1990s; but, rather than being entirely novel, they were features also of an earlier green activism (Scott 1990: 14).

Indeed, this brief survey endorses Steinmetz's sardonic comment that a 'cottage industry has grown up around the project of proving that the new social movements were really not so new after all' (1994: 179). Many EF! (UK) activists are hostile to any attempt to label their project as an instance of 'NSM' mobilisation:

> The 'new society' worships all that is new. Buy new Ariel automatic. Buy new activist – fully body-pierced for a limited period only...nothing is truly new – with the exception of the scale and complexity of the problem. Our struggles are recent battles in an old war.
>
> (*Do or Die!* 1997 (6): 70)

Papadakis, in turn, argues that examination of the history of environmental action is vital for academic understanding as well as activist commitment:

> Proponents of environmentalism in contemporary society have often ignored the efforts made by previous generations to address similar problems....The historical approach will assist us in analysing patterns of beliefs and behaviour that occur and recur. It will also enable us to identify what is or is not distinctively new about environmentalism in the latter part of the twentieth century.
>
> (1993: 45)

An historical approach can help to explain why protest occurs – for example, examining 'why, seemingly from nowhere, a popular women's peace movement sprang up in 1980–81'. Liddington suggests that

> it is necessary to glance at the unpopular peace politics of 1975–80. In retrospect it becomes clear that Greenham was presaged by tiny women's peace

> groupings, by imaginative direct actions against nuclear power, and by debates about feminism and nonviolence.
>
> (1989: 203)

Road protest is far from new, but has waxed and waned since the time of the introduction of the internal combustion engine. Green politics has an even longer history, and modern anti-road actions have clearly been fuelled by the more general expansion of green activism. Tactics have been reinvented and different green campaigns have cross-fertilised the forms of action. Yet ever since the arrival of the first Benz motorcar, greens have fought for and, apparently, won influence – only to find their words used as an excuse for further road building. For example, Tripp utilised environmental themes to argue for the segregation of road vehicles and pedestrians, suggesting that city precincts should

> cease to be maelstroms of noise and confusion, and become companionable places, with an air of leisure and repose; such streets will provide a real promenade for the town dweller and a rest for jaded nerves. We shall be getting back to Merrie England.
>
> (Quoted in Wistrich 1983: 66)

Such discourse has legitimised motorways and the great car economy via Buchanan's work in the 1960s and successive demands for bypasses in the 1980s and beyond. Demands for economic growth and cyclical recession, leading to large PSBR deficits, have in turn governed the pace of the motorway programme. Continuities are evident, but the pace of economic accumulation, the tendency to recycle environmental discourse as something very different, and the scale and nature of environmental protest have all accelerated through the 1990s, as the next two chapters will show.

3 The origins of Earth First!

> Someone phoned me up and said: 'There is this whole group of eco-fascists coming to Reading and wanting a platform. What do you think we should do about it?', and I just thought 'Oh my God!' I was sharing a stage with some of them...[and] it wasn't doing very much for my feminist credibility....So I was involved in that, and that kind of woke me up to some of the things they were saying and all the ideas of...what's his name, Dave Foreman...on population and AIDS, you know, people are being allowed to die in Africa and famines because it is part of the natural cycle, and it just turned me absolutely cold.
>
> (Tilly in interview)

Introduction

In 1980, five environmental activists took a trip into the Pincate Desert in Mexico. Venting collective frustration at what they perceived as the failure of the US environmental movement, they decided to create a new group (Lee 1995: 25–37; Manes 1990: 66–70; Scarce 1990: 58–62; Zakin 1993: 132–4). According to one of the five, 'Foreman called out, "Earth first!"....Roselle drew a clenched-fist logo, passed it up to the front of the van, and there was Earth First!' (Scarce 1990: 61).

Eleven years later, two students sat in their college refectory in Hastings, east Sussex, eagerly reading a copy of the US *Earth First! Journal* ('Mary' in interview). Disillusioned after participating in several green groups, the two decided to create an EF! movement in the UK (Torrance in interview).

These two accounts, despite their romanticised elements, illustrate how a tiny number of individuals launched new green networks advocating direct action. In the case of EF! (UK), it is instructive to consider how two young activists initiated a new network and helped launch in the UK the direct action anti-roads movement of the 1990s.

Resource mobilisation and repertoires

Resource-mobilisation theory (RMT) examines movement emergence and strategy by focusing on the ability of activists to gather appropriate resources, including not only finance but less tangible cultural assets such as tactical repertoires. RMT, which emerged in the late 1960s in the USA, attempts to explain the rapid growth of professional pressure groups or social movement organisations (SMOs) (McCarthy and Zald 1973, 1977; Oberschall 1973). At the time American political scientists and sociologists were tempted to explain direct action movements as instances of madness in an otherwise sane and stable society. In contrast RMT saw protesters as 'rational' agents who sought to maximise personal benefit via their actions. Emphasis was placed on the need to lower the 'costs' and increase the 'benefits' of movement participation, so as to kick-start active campaigning. Movement emergence was explained in terms of changes in the 'social resources available' (Jenkins and Perrow 1977: 250), with wealthy external benefactors seen as necessary to the creation of effective protest.

Direct action has been seen as a resource-efficient and cost-effective tactic (Taylor 1992). Groups with limited resources and limited input into powerful political networks may use direct action to increase the cost in moral or even directly financial terms of opponents' decisions. The British Animal Liberation Front (ALF), which influenced some EF! (UK) activists, has equally emphasised to vivisectors and other animal abusers the economic costs of sabotage:

> If you go and damage a laboratory they [the laboratory owners] have to pay to put it right and to install extra security measures (because often they won't get insurance unless they put in extra security). This money often comes out of their research budget and would [otherwise] be spent on experimentation. A lot of people criticize the fact that damage is done, saying that property is sacred. I think it is important to point out that damage to property does save animals.
>
> (Lee 1983: 10)

Many EF! (US) activists argue that ecotage 'can actually prevent destructive activity underway – driving the worst Earth destroyers right out of business – erasing their profits by slowing their work and destroying their tools' (Taylor 1991: 263). Road protesters in the UK have described their activities in similar terms:

> In it is a man, standing in the one fork of branch that the tiny tree has. He's been there for six hours. And he stays until they go. The tree lives for a few

> more days, and to finish their work they'll have to come out here, chainsaws, police. security, the full kit, on another day just for this one tree that's only just big enough to stand in. Difficult. Lengthy. Expensive.
>
> (Merrick 1996: 33)

Despite attempts to apply RMT assumptions to radical environmental action, a number of criticisms can be made of the model, as outlined above. First, naked self-interest fails to explain involvement in movements such as EF! (UK) that challenge the very concept of economic rationality. Second, green movements tend to have an activist base wider than just those individuals threatened by local environmental problems or individuals seeking employment as 'political entrepreneurs'. Third, the relationship between 'donors' and 'movements' is inadequately explained by RMT: it seems unlikely that the aims of groups committed to far-reaching change would attract support from elites. Indeed, it has been argued that 'organizations endure...by abandoning their oppositional politics in return for resources supplied by an elite' (Piven and Cloward 1977: xi). Such 'realist' political strategies allow movements to survive only by shedding their fundamental demands (McAdam 1982; Piven and Cloward 1977, 1992).

Yet resource considerations *can* influence a movement's progress. Interview accounts suggest that the founding EF! (UK) activists sought to mobilise resources to facilitate the movement's emergence. Although the requirements had been relatively modest, failure to mobilise capital, cultural and physical, may in part explain why earlier attempts to mobilise EF! in the UK failed (see pp. 45–6).

Pre-existing networks have been seen as vital to movement mobilisation. Not only are such networks recruiting grounds (see Chapter 4) but they may ease the mobilisation of resources in general. While such 'donors' from extant networks may share at least some goals with a new movement, 'donations' are never 'free' and may influence the issue-focus, broad ideological assumptions and forms of action taken by that movement.

Cultural resources, including political beliefs, vibrant symbols such as EF!'s monkeywrench and tomahawk logo, or tactics like 'tree-sitting' and 'digger-diving', are increasingly seen as vital to movement mobilisation. The idea that sets of tactics, described as 'repertoires of action', can be learnt, adapted and used for resource-deployment activities has been widely applied to social movements (Kriesi *et al.* 1995: 119; Roseneil 1995: 99–100; Tarrow 1994: 31–47). Repertoires are made up of tactics already familiar to activists, having been derived from existing movements. Apparently simple methods, such as the barricade, protest camp or strike, are borrowed from earlier protest movements: protesters rarely start from scratch.

Earth First!'s origins in the USA

All of EF! (US)'s founders had participated as professional activists in existing environmental networks and conservation SMOs such as the Sierra Club and the Wilderness Society. Indeed, EF! (US) is said to have initially 'looked a lot like the old Wilderness Society' in terms of its participants (Zakin 1993: 142). EF! (US) boasted of its lack of financial and physical resources, claiming that it did not even own 'a photocopier machine' (Zakin 1993: 145). Eventually the EF! (US) newsletter became a tabloid newspaper, and modest office facilities had to be found. Both activists and finance were mobilised by exploiting existing networks. In particular early contact was made with influential sympathisers such as FoE's founder David Brower and, of course, Edward Abbey (Zakin 1993). The first large-scale EF! (US) Gathering, the 1980 Round River Rendezvous, was used to mobilise and recruit activists from the Wilderness Society (Zakin 1993: 142). Continuing recruitment attracted increasingly diverse social-movement activists and anarchists, gradually changing the character of EF! (US) (Lee 1995: 119).

Cultural resources were also derived from existing networks. The concept of 'wilderness' has a lengthy history in the US via the writings of Sierra Club-founder Muir and former park ranger Leopold (Manes 1990: 23–4; Zakin 1993: 63), plus Abbey's. Equally EF! (US) was able to draw upon repertoires of environmentally motivated sabotage that had been refracted through literary sources. Abbey based *The Monkey Wrench Gang* (1973) on 'a group known as the Eco-Raiders, who...used unconventional and illegal tactics to slow the growth of the suburbs of Tucson, Arizona' (Lee 1995: 26). During the 1970s other 'eco-saboteurs', for example 'the Fox', attacked the offices of large corporations and dismantled pylons carrying nuclear-generated electricity (Mowrey and Redmond 1993: 73; Nash 1989: 191; Zakin 1993: 59).

Repertoires and symbols were derived from other US movements. Mike Roselle had participated in NVDA against the Vietnam War (Scarce 1990: 164), using a repertoire that was later adapted and widely practised by EF! (US) along with more controversial forms of 'ecotage' (Lee 1995: 72; Manes 1990: 87). The clenched fist he drew as part of the EF! (US) logo can be seen as a symbolic appropriation of the symbol of the US Black Power movement to whose influence Roselle had been exposed while a member of the Youth International Party.

Initial actions were humorous, symbolic, provocative and, ultimately, threatening. In April 1980 Foreman helped to erect a mock memorial to an Apache leader who had attacked a mining camp in the New Mexico desert, seeking to appropriate him as an 'eco-warrior'. The plaque stated:

> This monument celebrates the 100th Anniversary of the great Apache chief Victorio's raid on the Cooney mining camp near Mogollon, New Mexico, on April 28, 1880. Victorio strove to protect these mountains from mining and

other destructive activities of the white race. The present Gila Wilderness is partly a fruit of his efforts. ERECTED BY THE NEW MEXICO PATRIOTIC HERITAGE [*sic*] SOCIETY.

(Manes 1990: 73)

In March 1981 a demonstration was held at Glen Canyon Dam, Arizona. The dam had been built after a compromise by the Sierra Club and had become for EF! (US) activists a symbol of the weakness of conventional environmental organisations. In Abbey's novel the monkeywrenchers plan to demolish the dam and succeed in blowing up bridges, burning bill boards and wrecking bulldozers (1991: 15). On this occasion Abbey watched as the outrageous goal of his novel was symbolically completed: while other demonstrators distracted police, a 100-metre polythene 'crack' was unrolled from the parapet, creating for a few seconds the illusion of dam destruction (Manes 1990: 6).

While such actions indicated an antipathy to conventional environmental campaigning and an apparent commitment to violent action, EF! (US) actions have remained largely symbolic and non-violent. The movement has never, for example, attempted to actually demolish dams; nor have activists physically attacked opponents in the same way as had their Apache hero or Blott in Sharpe's very English ecotage novel. Despite the instructions given in Abbey's novel (see p. 3) neither have explosives ever been used: EF! (US) was not to be a green militia. Increasingly, though, EF! (US) adopted repertoires of both covert sabotage and open NVDA.

The combination of such different repertoires led to practical and political problems. Some activists felt that publication of sabotage manuals, such as *Ecodefense* (Foreman and Haywood 1993), and calls for illegal action placed individuals practising NVDA at risk of reprisals from opponents, including state authorities (Lee 1995: 134). 'Tree-spiking', which would wreck chainsaw blades or mill equipment, was condemned by some in the movement as dangerous to timber workers (Bari 1994: 271–82; Rowell 1996: 152–3).

Some founding EF! (US) activists initially advocated a set of conservative naturalist beliefs, drawn from a misanthropic reading of deep ecology (Foreman 1991: 55–8; Lee 1995: 61). Population growth was seen as the most important source of ecological crisis, and racist statements were issued in support of restrictions on Hispanic immigration into the US (Lee 1995: 106–11). Such sentiments were criticised not only by social ecologists and other radicals in the US green movement (Bookchin 1988; Bradford 1989: 51) but increasingly by a 'social-justice faction' within EF! (US) (Lee 1995: 109).

A minority of EF! (US)'s early activists boasted of their conservative affiliations, love of the American flag and 'redneck' background (Lee 1995: 32). Yet from the very earliest days the 'social-justice faction', rooted in wider social-movement activity including anti-nuclear power, feminist, labour and peace networks, grew within EF! (US) (Lee 1995: 120–2). This group, whose principal advocates included founding activist Roselle and Californian labour agitator Bari, argued that opposing social injustice has an importance equal to that of working for wilderness protection. In turn they saw environmental and social problems as rooted in social rather than exclusively 'natural' causes (Bari 1993 and 1994). In 1990 Foreman resigned from EF! (US) because of his perception of the increasingly dominant influence of natural over social issues. He had earlier stated: 'for a group more committed to gila monsters and mountain lions than to people, there will not be a total alliance with other social movements' (Lee 1995: 57). Despite this lengthy factional dispute, EF! (US) established a national network and remains active today (Nyhagen Predelli 1995: 124).

Thus by the time the founding EF! (UK) activists made contact with EF! (US), the latter had created a North American network, suffered internal divisions, become a target of state and counter-movement forces and was notorious both for the early ideological conservatism of some of its activists and for its continuing practice of disruptive and more symbolic direct action. Roads that brought people and vehicles into areas of wild nature were opposed by EF! (US) and the network has supported anti-car actions in major US cities. Yet the network has never been part of a mass anti-roads campaign, as in the UK, and remains, despite the influence of activists like Bari, more concerned with mountain grizzly bears than with car culture.

Failed Earth First! mobilisations in the UK

Chris Laughton, a physics graduate, had attempted to launch EF! (UK) in 1987 (Laughton, Noble and 'Mix' interviews). Laughton had worked as a Greenpeace International volunteer in the USA during 1986, but became disillusioned: 'I was attracted to Greenpeace's direct action [but] I became very, very cynical about the high wages they paid the directors.' Other volunteers put him in touch with EF! (US) in New England and he subscribed to the *Earth First! Journal* on his return to the UK:

> I found it very inspirational...it was direct action with a capital 'D', it was ecotage, it...sounded very exciting and interesting and it then made sense that Greenpeace was a mainstream organization...whereas Earth First! was a partly subversive organization and there was something interesting and attractive about that....I felt there was a place for it in the UK.
>
> (Laughton in interview)

Laughton imported fifty copies of *Ecodefense* (Foreman and Haywood 1993) which he sold to radical bookshops. He added his name and address to each copy as an EF! contact; he also approached green journals including *Green Line* and *Green*

Anarchist. He promoted EF! through the *Green Anarchist* network and was listed in EF! (US) publications as the UK contact. Yet after several months of activity he felt that he had failed to 'get the ball rolling....I failed to meet anybody although it wasn't for lack of trying....I ended up in a hippy commune for three years...changing my life to get closer to the Earth'.

Collie, a Scots activist, organised an EF! network in Scotland which staged demonstrations on peace themes, but by 1991 his group had ceased to be active (Laughton in interview; Smith 1987: 3). Finally, activists from south-west England, like Laughton, perceiving EF! to be an 'underground' body practising a repertoire of covert sabotage – mainly attacks on JCBs and other earth-moving equipment – were active in the summer of 1991. They were, however, reluctant to organise openly and found it difficult to publicise their illegal actions or make links with other 'ecoteurs' active at the time (Wall 1991b).

Earth First! Hastings

A more enduring EF! (UK) network was created by two further-education students, Jake Burbridge and Jason Torrance, from Hastings, east Sussex, in the spring of 1991. Previously active in FoE, the Green Party, Greenpeace and the peace movement, they had become critical of existing green networks and sought a new approach to environmental activism (Torrance in interview). Burbridge bought a copy of the book *Deep Ecology* (Devall 1985) and found an address for EF! (US). He wrote to them and before too long became their UK contact.

Both individuals showed an enthusiasm for what they perceived to be EF! (US)'s belief in deep ecology and its repertoires of action ('Mary' in interview). Yet their earlier involvement in the broad green movement family continued to shape their approach to action. Thus the first EF! (UK) action, a blockade of the Dungeness nuclear power station, copied existing peace movement and anti-nuclear power tactics. Indeed, in reference to his later involvement at Twyford, where he sought to transfer a repertoire of action derived from the peace movement, Torrance noted: 'We were known as the direct action people and we'd bring in our skills from the peace movement.' Participants were drawn from local peace and anti-nuclear power networks:

> Brighton Peace Centre and some other Sea Action [peace group] people came along and we worked as well with DASH, which is Dungeness Action Society of Hastings, and we organised a sixty or so strong demonstration.
>
> (Torrance in interview)

The action was framed from a deep-ecology perspective, with a press statement observing that the Dungeness area contained 'the best example of a cuspate foreland in the world...home to over 600 species of flora and fauna, some of which are rare' (*Green Anarchist* 1991 (28): 24). Dungeness, a remote area of shingle, famed for its lighthouse, is perhaps the landscape closest to 'wilderness' in south-east England.

Efforts were made to sustain a Hastings EF! group, but it remained a tiny and ephemeral body. Burbridge noted: 'We've held a picket against a skip hire company which is cutting down woods. There's been fly-posting and painting on them. We only have ten people in the Hastings group' (*Green Anarchist* 1991 (26): 15). Torrance stated: 'realistically, it was Jake and I [who] tried to get lots more people in.... We had bigger meetings and a fair amount of people had come but I guess, looking back on it, it wasn't meant to be like that.' Increasingly, national and international concerns were to absorb Burbridge and Torrance's time and mobilising expertise.

An early contact was George Marshall, a former libertarian right-winger; his politics had been transformed while taking a degree in sociology in the late 1980s, and he was becoming alarmed by global environmental problems (Marshall in interview). The Conservative Students, of which he had been a member at university, was a right-wing and often eccentric student section of the Conservative Party. During the 1980s the Conservative Students had supported the right-wing guerrillas in Angola and Nicaragua, were violently hostile to South Africa's future president Nelson Mandela, and had flirted with policies of legalising drugs, soft and hard.

Marshall, to a greater extent than had the conservative founders of EF! (US), and Laughton, who also had sympathised with the right, moved to the left as a result of his green conversion. Initially, Marshall was active in the Australian rainforest movement and worked with John Seed. Seed in turn had taken part in EF! (US) actions in the early 1980s (Taylor 1995: 18; see Chapter 8). Inspired by Foreman, he built strong links between the Australian direct-action movement and EF! (US). Seed influenced Marshall, as well as Australian activists who came to Britain:

> His approach was completely grassroots. He lived on the dole or he...scraped together money here and there. He'd have a van [and] he'd just trundle round Australia. He'd turn up in a town, and he would say, 'Hey! Public meeting, public meeting, come along.' He showed his slides and he'd say rainforests were the womb of life, home to half the world's species of plants and animals, and [he'd] give his little rave, his hippie rave, and sing a few songs and say, 'Right, what are you going to do about it?' There would be a few people who would hang on after the meeting; he would talk to them and they would become the core of the rainforest action group. And after that, he would feed them and support them – it is very much an Earth First! approach.
>
> (Marshall in interview)

Returning to the UK in 1990, Marshall and Shelley Braithwaite, an Australian activist, worked to create a UK rainforest movement. They hoped to use Seed's organisational approach and repertoire of action, but the attempt to apply ways of doing things learnt in Australia met with little initial success:

> I wanted to do direct action, [but] I couldn't find anybody to do it. About this time John Seed wrote to me…saying: 'I noticed off the latest Earth First! contact list an address in Britain. Why don't you get in contact with them?'
>
> (Marshall in interview)

Having contacted the two Hastings activists via the US *Earth First! Journal*, Marshall struck up a strong affinity with them:

> Jason describes it as love at first sight…it was people who thought the same way, who had the same perspective, who…[agreed on] the scale of the issues and what was needed to move it in the same way, and I was absolutely ecstatic after all this time of trying to find people who could do it.…From the very beginning, as I recall, we started talking about working together quite seriously.…They were very clear and focused. It was, like, 'Fuck all this namby-pamby stuff, we want some action' – and that's actually just what I needed.
>
> (Marshall in interview)

As Torrance recollects:

> And we also tracked down George Marshall [of]…the London Rainforest Action Group, and he was, again, another upper-class environmentalist. I was really becoming aware that the movement was just full of upper-class public schoolboys, at that point…but he was really open and…welcoming to me, which was a real change. He came straight from Australia, and had been banging his head up against a brick wall trying to get local grassroots' direct action off the ground…again, just like me and Jake. Meeting and working together, we really clicked. We both had things to offer, and in a way we both needed each other to get things going…he was really into getting things going in this country.
>
> (Torrance in interview)

Marshall provided EF! (UK) with an issue-focus – rainforests – and a real ability to mobilise resources. He had strong international contacts and worked for the influential magazine *The Ecologist*. His contact with *The Ecologist* provided indirect links with the founding editor's billionaire brother Sir James Goldsmith, who had recently committed himself to increasing financial and political support for environmental projects (Fallon 1991: 473). Goldsmith's finance for rainforest action in the UK was used to help fund early EF! (UK) projects (Torrance in interview). Marshall was thus able to help tap into existing green networks to supply the emerging EF! (UK) network with finance, and physical and cultural resources. His experience, and that of Australian activists such as Braithwaite, provided an action repertoire, derived indirectly from Australia, that complemented the peace and anti-nuclear power repertoires of NVDA used by Burbridge and Torrance.

In June 1991 Marshall helped create an Oxford Rainforest Action Group (RAG), which within a few months became Oxford EF!, as well as a London RAG; he had also organised a number of actions. Burbridge and Torrance worked equally hard to make links with existing UK green networks, sending circulars to student green groups, peace-movement organisations and Green Party branches.

An interview with Burbridge and Torrance in *Green Anarchist* was read by Karen Noble. Her contact with them led, eventually, to the formation of an EF! group in Littlehampton, west Sussex. Noble is known as one of Britain's leading fruitarians, and, along with her advocacy of a pure raw-food diet, she strongly supported radical animal rights and green causes. Having made contact with Burbridge and Torrance, she encouraged them to make links with a tiny network of deep ecologists and EF! (US) sympathisers, such as George French from Northumbria who wrote for the US *Earth First! Journal* and Bob Finch who published anti-car polemics. Noble had also been in contact with Laughton, who in turn visited Burbridge and Torrance during 1991.

'Active' participation was important also as a way of gaining new contacts. Taking part in a second anti-nuclear power action, Torrance noted:

> Jake and I had done an action up at Sizewell on the Suffolk coast...to try and prevent a reactor heart coming into Sizewell from Germany...[via] Lowestoft harbour, and we went out with Sea Action. Jake and Angie Zelter ended up chaining themselves onto the vessel that was carrying the reactor heart, and I was in the inflatable, buzzing around, having a whale of a time. And that's how we met Angie....She's a really strong woman, and Jake and I were just really impressed by her....You can imagine, at the time we were sort of greenhorn activists and here was Angie who had been involved in Greenham and, like, she was just 'Oh, no! The reactor heart is in there. I feel we've got to do something. Let's go and break in.' And, like, she's in her mid-40s and a real power house, and she's got real crazy fucking eyes.
>
> (Torrance in interview)

Zelter was to influence the emerging EF! (UK) to participate in its first international campaign; she also provided links to other activist networks. As an activist in the peace movement, she had sought to develop new repertoires of action:

> 'Snowball'...the enforce-the-law campaign, came out of my direct action experience of Greenham and wanting to carry Greenham home, and also from trying to make direct action more accessible to ordinary people. So it was devised really to make it very easy to confront the military. Sort of all you had to do was cut one bit of chain-link fence with a hacksaw-blade. That was technically criminal damage [but] we would argue that nuclear weapons were in breach of the genocide act. It was a way of finding something for people to do, for expressing politically through the court system their opposition to nuclear weapons.
> (Zelter interview)

Zelter formulated, or else adapted, a series of 'legal–symbolic' NVDA acts. Initially she developed Snowball in the peace movement; later she helped develop for EF! (UK) and other rainforest campaigners the practice of 'ethical shoplifting'. Later still, in the mid-1990s, she become active in the Ploughshares movement, using similar acts of ritualised illegality to oppose arms sales to Indonesia (Pilger 1996b: 5). Then 1998 saw the genetic Snowball Campaign, with activists up-rooting genetically engineered crops. In all these campaigns direct action is legitimised and utilised with reference to existing legal forms. Typically, timber taken, allegedly illegally, from areas of rainforest is seized by 'ethical shoplifting' EF!ers and other rainforest campaigners. Such actions were seen as humorous and ironic, but non-threatening – in contrast to EF! (US)'s early acts, such as the 'cracking' of Glen Canyon Dam. Because of their tokenistic or iconic approach to direct action, their humour and their compliance, however tenuous, with legal structures, Zelter's tactical adaptations were seen as providing an easy introduction to direct action for new participants (Tilly in interview).

Rainforest campaigning

During the 1980s and early 1990s the image of a burning rainforest had become a powerful symbol of global environmental destruction (Hannigan 1995: 45; Pearce 1991: 183), and rainforest actions were to provide EF! (UK) with an ideal issue-focus with which to mobilise existing green activists. In July 1991, Burbridge and Zelter flew to Sarawak in Malaysia, where they were arrested after an international rainforest action and imprisoned for two months (Burbridge 1991: 10; *The Times* 19 July 1991: 12). The action, held at the invitation of EF! (US) activist Jake Jagoff, was in support of the (local) Penan campaign of civil disobedience against rainforest destruction (Gedicks 1995: 94–8).

Marshall made use of his access to finance, and links with the peace movement were mobilised, to resource an EF! (UK) office to support the Sarawak campaign and other early actions. Torrance had worked as a volunteer for the Gulf Peace Team, a group with offices in Stoke Newington, who had attempted to halt the 1991 war by creating a peace camp in the Iraqi desert.

> Things from an organisational viewpoint peaked in this country when we had a meeting at the Gulf Peace Team Offices at 7 Cazenove Road, Stoke Newington.... We faxed Jake Jagoff in Australia and decided it was going to happen. Angie and Jake would go out there and I'd stay behind to do support work along with George... We said: 'You know we need an office.' I'd already at that point bargained to have a table, a desk in the Gulf Peace Team Offices.... I'd been working in Number 9 which was really nice, [a] really plush office two doors up. And George said: 'Money is no problem. We've got the money, we just need to find a location.' I, sitting in Number 9, said: 'I know a perfect location.' [DW: This was Goldsmith money?] 'Yes.'
>
> (Torrance in interview)

While the Sarawak experience increased the initial activists' commitment to the rainforests issue, and raised the profile of the network, it increased hostility between EF! (UK) and established environmental pressure groups. FoE, for example, were strongly critical of the action, arguing that it helped the Malaysian government to claim that 'protests against rainforest destruction in Sarawak are generated by "imperialist" outside agitators' (Lees and Bourn 1992: 2). Burbridge (1991: 10) noted: 'In our absence from Britain we have been tried and convicted by the mainstream environmental groups. They have convicted us of a crime they themselves could never be accused of: action. With friends like these, the Earth doesn't need any enemies.'

While Burbridge and Zelter were in jail, EF! (UK) targeted the July 1991 G7 – heads of government – Conference. Marshall described the event as

> a primary opportunity to confront... the heads of major scumbag countries in the world... [and] it was happening in London. I can't remember how many, like, 10,000 media were there, or something; it was a real chance to get a message out, a real chance to confront them directly. Nobody did any actions at all except for us. Nothing.
>
> (Marshall in interview)

Bruno Manser, a Swiss rainforest activist contacted by EF! (UK), climbed a lamppost outside the conference (*Independent* 18 July 1991: 1). Using a press pass, Marshall gained entry and launched a verbal assault on Prime Minister John Major.

Protests against participation in rainforest destruction were organised outside the Japanese Embassy and at the headquarters of Japanese firms. In October 1991 the Malaysian Tourist Office was occupied and sawdust dumped on the floor. One month later a banner was dropped from Australia House in protest at plans to log Australian forests. These actions succeeded in generating strong media images;

they also widened access to activist networks within the green movement. Such access in turn allowed the introduction of mass NVDA repertoires, transferred from both the UK peace and anti-nuclear power movements and the Australian Rainforest Campaign, which commanded the support of hundreds rather than dozens of activists.

The first 'mass' action occurred on 4 December 1991: EF! (UK) attempted to prevent the *MV Singa Wilstream* docking at Tilbury, on the Thames Estuary, with a cargo of rainforest timber from Sarawak. This intervention, based on an Australian model of 'ship actions', was preceded by an extensive mailing to green movement contacts. Sea Shepherd, the Greenpeace militant splinter-group, supplied the boats used in the action. Although it failed to prevent the ship docking, the action was seen as helpful in building an activist base for NVDA from within the green movement and in generating greater media attention:

> How could stopping an international port not be big news? So FoE came in on it with their huge inflatable rubber chainsaw. That, of course, got in all the press photos and, you know, it was really great to sit down in a huge line of people all [with] linked arms, blockading the gate, people...from WWF, Survival International, FoE, local groups, EF! local groups, RAG local groups, Green Party, you name it. There was probably 150–200 people there....Tilbury was really the first ambitious action we had done in this country.
>
> (Torrance in interview)

A second maritime action was held in Liverpool, in the spring of 1992, where 400 activists occupied the docks. This action was organised to coincide with the end of a National Green Student Network (GSN) Conference in the city, so as to maximise participation. The organisers proclaimed in a mailing to other EF! (UK) groups:

> We plan to disrupt the docking of the ship, by being present on the river, on the berth side and at the entrance to the dock....The most threatened rainforests are those of Indonesia and Malaysia. These are being destroyed at a rate of 13 square kilometres per day. If this continues, the Rainforest of Sarawak (Malaysia) will be exhausted within eight years. The Straits of South East Asia, a graveyard of what was once Rainforest, which has been entirely logged, is used as a storage depot of Malaysian hardwoods. At least two shiploads of tropical hardwood reach Liverpool every month, from Indonesia.

In May 1992 over 200 activists occupied a timber yard outside Oxford, forcing it to close for the day. I arrived in a mini-bus with students from Cambridge

University, where I had been visiting Graham Jeffrey, a fellow green revolutionary. The afternoon was spent in training workshops, the evening in a woodland camp. We were woken early the next morning, and marched onto the site at 5.00 a.m. The police presence was small, presumably because they had not expected us to arrive so early. Within a few hours we had occupied the whole site. Much gentle mayhem resulted, with labels being torn from timber and vast free-form spider webs of wood piled up. Angie Zelter ritually sawed a piece of wood on behalf of a Malaysian tribe. Given the amount of ecotage of a non-ritual kind that I had witnessed, the iconic splendour of this act rather escaped me. Relations with the workers were surprisingly good, despite some initial friction. Some protesters played football with one group of workers. At midday the workers marched out of the gates, followed by EF!ers. Davy Garland, among others, had negotiated a full day's paid holiday for the workers. It was, in short, a very impressive action. A sting in the tail for the firm, Britain's second-biggest importer of tropical timber, occurred several weeks later when, after a tip-off from protesters, Customs and Excise officers seized twenty-eight tonnes of alerce, a protected and endangered Chilean softwood. In June 1992 a similar EF! (UK) action occurred at a Rochdale timber yard.

By 1993 such actions had ceased: micro-programmes, in the form, for example, of Zelter's ethical shoplifting, replaced mass NVDA. Not only had the roads issue grown in importance but there is evidence that police activity was making dockside and timber-yard actions less effective as means of disruption and in generating media attention:

> One thing Earth First! were really crap at was this idea of running continuous campaigns. They were very good at...one-off demos but it [*sic*] wasn't sticking with it. I caught the end of the rainforest campaign, and there was this thing against Scott Paper in Northfleet, and the police [did] what they always do – they close the works for the day, stuff loads of people there and just expect you to sit outside and get bored. It is boring for the demonstrators. It is boring for the papers – hopefully nobody comes again...and the other 364 days of the year it is business as usual for the Earth-rapers.
>
> ('Mix' in interview)

The failure to maintain rainforest mass actions disappointed some:

> The only place in the world that this had ever been done before was in Australia....George and Shelley had really brought this idea over with them...and they still talk about it now, but it never really took off...the whole roads thing overtook them.
>
> (Torrance in interview)

It seems clear that the rainforests issue had a far stronger resonance in Australia, where activists were familiar with forests actually threatened, than in the UK, with its very different natural environment. Rainforest protests allowed EF! (UK) to mobilise hundreds of activists, but wider participation, drawing not just on core green movement radicals but on more extensive networks, was to come with the new focus of anti-car and road-protest action.

Divisions

During EF! (UK)'s first year of mobilisation, political divisions grew as new activists challenged the initial issue-focus, the political assumptions and the repertoires advocated by Braithwaite, Burbridge, Marshall, Torrance and Zelter (a view supported by Garland, 'Mix', Molland, Noble and Torrance, respectively, in interview). Typically, Davy Garland (see p. 6), an anarchist, hunt-saboteur and former Communist, had taken part in European militant anti-nuclear actions and animal-liberation activities:

> I was involved in the big one at Borsock, at the power station in central Holland, and we had a three-day blockade...we basically shut them down for three days. They sent riot police, but what I liked was, on one side we had all the helmeted autonomen with their shields and truncheons, who basically took the gates off, and down the road we had all the peaceniks chaining themselves, but the two, the two were [very] much together. So through this I made the connection [to] direct action.
>
> (Garland in interview)

Garland, who created Mid-Somerset EF!, the second EF! (UK) local group (Torrance and Burbridge's was the first), believed that direct action should link environmental and social issues unequivocally. He stressed the importance of mobilising community support for local issues, as well as international issues like rainforests. He also felt that direct action should be physically disruptive rather than aimed at generating media images. As Torrance noted: 'Davy was more your coming-from-the-class perspective, and [the] South Downs' were coming more from [a] hunt–sabby–squatty kind of perspective.' Garland observed: 'I believed in...revolutionary politics, community politics plus animal liberation.'

Such activists sought to legitimise sabotage as part of EF! (UK)'s action repertoire, believing that these approaches were licensed by EF! (US)-founder Foreman's support for ecotage, the publication of *Ecodefense* and the slogan 'No compromise in defence of Mother Earth' (Nash 1989: 191). Repertoires of covert sabotage practised by the ALF were an obvious source of inspiration for such activists, who in turn used animal liberation networks to recruit others sympathetic to their approach into EF! (UK):

> A good way-in was to approach the hunt-saboteurs, because they are already into sort of direct action and it is quite easy to show them the links....If you are going to save the fox from the hunt, why let it be killed by a contractor?...From that, Plymouth EF! was set up....The thing that really appealed to me initially was the deep ecology that I saw as an extension of the animal liberation idea.
>
> (Molland in interview)

In contrast, other EF! (UK) activists, particularly those from peace-movement networks, were uneasy about covert repertoires and links with animal liberation militants. They feared that such an approach might draw police attention to the movement, and it conflicted with their own cultural assumptions inherited from previous activity, in particular an unambiguous commitment to total non-violence. Tilly noted her experience of previous debates on sabotage: 'It was something that had never been done in the peace movement....I mean, there'd been whole big meetings at Greenham about whether or not we should cut a fence. You know, I remember that as being a big, big thing that happened.' Such 'moderates' believed in open forms of NVDA, feeling that if sabotage occurred its focus should be symbolic and that activists should take responsibility for any damage done. In contrast 'militants' felt that covert sabotage and self-defence were legitimate elements of EF! (UK)'s action repertoire: one women activist told me that 'violence to machines is thoroughly welcomed and blessed'. It is only a minor simplification to suggest that those activists drawn from the peace movement saw EF! (UK) as a means of promoting mass NVDA of a largely symbolic form, while those from an animal-liberation background regarded EF! as a vehicle for more militant tactics.

The wider ideology of EF! (US) was interpreted in equally dissimilar ways by different activists within EF! (UK). To some, deep ecology was consistent with or even identical to existing assumptions held within the UK's green movements (see Molland, above). To others, the deep ecology of EF! (US) was perceived to be racist, misanthropic, sexist and in conflict with prior green assumptions.

Such debates were accelerated by the EF! road-show in the spring of 1992, when four EF! (US) activists toured the UK, making speeches, showing slides of campaign activity and playing folk music. Two of the activists, Jake Jagoff (who had made early links with Burbridge and Torrance) and Mike Meese, emphasised wilderness preservation, population control and anti-humanist views. To many of the activists who helped found EF! (UK), such views were controversial, even repellent. At one meeting in Reading, anti-fascists, including individuals who later became leading EF! (UK) activists, picketed the road-show. I organised a Bristol University meeting for the road-show and was somewhat alarmed by phone calls from the Reading activists, as well as by the Malthusian views of

Meese. At Reading critics distributed a leaflet asking whether much racism was involved in EF!

> We got some very dubious answers from the people who were speaking there.... I can remember there was one very big, bearded, American bloke who pointed to me and said: 'Don't you fucking call me a racist,' so it was kind of exciting actually!
>
> (Anon. in interview)

Militants such as Garland worked to publicise the views of Judi Bari, and others in EF! (US), who sought to link deep ecology and social analysis (Bari 1992: 9). EF! (UK) women activists, some of whom had been active at Greenham Common Women's Peace Camp, were also critical of what they interpreted as sexist assumptions on the part of the US activists who toured Britain:

> I noticed these stickers that were on the filing cabinets, and one of them should have read 'Real Men Ride Bikes', but somebody had scribbled out the 'B' and put 'D', so it read 'Real Men Ride Dykes' [*sic*]; and the other one...should have read 'I'd Rather Be Monkeywrenching', which is the EF! term of ecotage, and somebody had scribbled out the 'r' so it read 'monkeywenching'...and I just went up the wall.... When somebody said 'Well, actually, I think that was done by the American Earth First! people'...I thought, yeah, that figures.
>
> (Tilly in interview)

> I don't know whether it came from the States, or whatever, but the first Earth First! office in London...was rather crude, rather druggy, drink-orientated and very male...really eco-warrior-type stuff.... Although I think that Earth First! in Britain is much gentler than America...it can be a bit offputting for women, unless you want to be associated with 30-year-old beer-swilling, you know, 'Great, go and be heroes' – that kind of hero stuff.
>
> (Zelter in interview)

Disputes over ideology and repertoire were paralleled in discussions of movement structure. Militants argued that both feminists and 'moderates' like Marshall wanted to create a hierarchical organisation (Garland in interview). Marshall rejected such charges, but argued that a clear and coherent democratic framework was necessary for movement growth (Marshall in interview).

In April 1992 EF! (UK) held its first national Gathering near Brighton. Sixty activists attempted to forge an identity for the movement and deal with such controversies. The feminist critique seems to have been accepted with relatively little dispute (Jukes and Tilly in interview). It was agreed that women-only space should be provided, both at EF! (UK) events and on 'actions', if women participants felt it appropriate. Several weeks later at the EF! (UK) blockade of the Timbmet timber yard near Oxford, Tilly organised a successful women's occupation of one of the site's entrances. In contrast to Greenham, most events remained mixed, and within a few months Tilly felt that the issue of sexism in the movement had receded. A women's page also appeared in the single-run issue of the EF! (UK) paper *Wild*, with a critique of 'political housework' arguing that women in the peace and environmental movements were often exploited by men, doing preparatory work but having little say in decision-making.

The issues of population control, alleged racism and the asocial nature of deep ecology were, however, largely ignored. EF! (UK) has never codified its ideology, unlike organisations such as the Green Party which spends much time at national conferences developing complex platform positions. Even the loosely anarchist networks Greenpeace London and *Green Anarchist* place far more emphasis on ideology than does EF! (UK). These omissions occurred despite the statement in the Spring issue of *Action Update* (1992): 'A primary aim of the Gathering was to define what exactly Earth First! is.' The term 'deep ecology' and the slogan 'No compromise in Defence of Mother Earth' are absent from the set of Aims, Methods and Principles agreed at the April 1992 Earth First! Gathering, which focused instead on repertoires of action:

AIMS

- To defend the environment
- To confront and expose those destroying the environment
- To realise a human lifestyle that exists in balance and harmony with the natural world and that has respect for all life.

METHODS

We will achieve these aims by:

- empowering individuals and groups to take direct and focused action against those destroying the environment
- networking information and contacts between action groups to facilitate the growth of a movement and encourage group autonomy
- raising funds for direct action campaigns and networking costs.

PRINCIPLES

The following are principles within which all groups have agreed to work.

1 Local groups should be empowered and supported to network their own campaigns and actions. Local groups organising an action can lay own

> ground rules for that action. Other groups involved should be informed of, respect, and comply with these rules.
>
> 2 We recognise that some people may feel moved to damage property. We neither condemn nor condone such actions. Such actions are the sole responsibility of the individuals involved.
>
> 3 We agree that violence towards living things is not a legitimate tactic in environmental campaigns. Therefore we will follow strict principles of non-violence when confronting the destruction of the environment.
>
> 4 We recognise the diversity of opinion within the environment movement and support the genuine efforts of other environmentalists in defending the earth.
>
> (*AU* 1992 (3): 5)

It can be seen here that no attempt is made to 'frame' EF! (UK) belief in terms of the assumed nature of environmental destruction. The emphasis is on 'methods', with support for direct action, a rejection of violence and support for 'group autonomy' or empowerment. Such concepts are left largely undefined. No mention is made of the possible conflict between principles 1 and 4, which are pluralistic, and 2 and 3, which reject both 'damage to property' and 'violence towards living things' as legitimate tactics for EF! (UK) groups. The statement was unenforceable and with changed circumstances seems to have been forgotten at subsequent Gatherings. There has been no attempt since to agree such a set of aims, methods and principles. Rather than being set in stone, the Brighton statement was soon forgotten.

The issue of violence attracted the most controversy at the Gathering and the narrative of EF! (US) as a violent extremist group was widely reproduced by the UK media (Charman 1992; Cohen 1992). While militants argued that neither sabotage nor self-defence constitutes 'violence', moderates argued for non-violence as a principle. Several days before the Gathering, activists in the EF! (UK) London office had been embarrassed by media questions in response to a communique that had attributed responsibility to EF! (UK) for sabotaging a Fisons peat-digging operation in south Yorkshire. During the Gathering a newspaper report was published containing an unattributed statement from an EF! (UK) activist according to which radical greens might carry out bomb attacks (Cohen 1992).

> At that meeting it was quite disturbing how many, like, dregs of the peace movement...all these ex-Greenhamoids and Cruise-watchers who [felt] they had got a new ground for direct action, by which they meant civil disobedience, and just immediately grafted all the bullshit that had polluted their mind onto this new movement....Jake or Jason had been conned by some reporter into going on about how bombings would soon be seen in the cause

> of environmentalism…so he was all terribly embarrassed by this. That was used as a stick to beat the militants with.
>
> ('Mix' in interview)

In contrast, moderates suggested that it was impossible to endorse sabotage openly:

> I think it turned into a kind of, you know, a very intense debate between the people who, how I regard anyway, between the people who really wanted a social-change movement and the people who just wanted [the] posturing, the people who…liked the idea…the image [of] eco-warriors, you see.
>
> (Marshall in interview)

After much debate 'it was agreed' at the Gathering that

> Earth First! would be split into two. On the one hand there would be an underground group, the Earth Liberation Front, which would do ecotage and all the embarrassing naughtiness stuff and, on the other hand, all the open civil disobedience kind of thing that would retain the name Earth First!…People were insisting that if there was going to be a split it shouldn't be a case of competition between units. They should be supportive, so there should be toleration by groups.
>
> ('Mix' in interview)

> For some, repertoires of sabotage were seen as intrinsic to EF! identity, [though] I didn't hear about any activities in the handbook [*Ecodefense*]. They were just publicly waving flags. Now for the complete movement, as far as I was concerned, the companies that were destroying our ecology, their insurance premiums had to go up, to make it more uneconomical for them to go ahead. Most of the stuff I have seen under the banner of EF! has been mass demonstrations.…I feel that EF! has become a bit abused as a term in Britain.
>
> (Laughton in interview)

The evolving character of Earth First! (UK)

By the time of the April 1992 Gathering EF! (UK) had created a national network of supporters who rejected (or, more accurately, ignored) right-wing interpretations of deep ecology and utilised a repertoire of NVDA.

Local groups seem to have been heterogeneously drawn both from disillusioned FoE activists and the more militant currents represented by Garland, Molland and South Downs EF! (Mercer 1995: 119). By April 1992, seven groups were listed by *Action Update*. Manchester EF! and Merseyside EF! had been created from Green Student Network university contacts. Merseyside EF! also functioned as a *GA* group. Oxford EF! had been a local RAG group. South Downs and South East London were influenced by the animal-liberation movement. Rapid growth occurred during 1992 and 1993. A May 1992 contact list showed 20 local groups and 11 additional regional contacts. The number of local groups grew to 50 by July 1994, after which date the number of 'named' EF! groups fell.

Many groups seem to have been very small. Local EF! (UK) groups were active in diverse ways. Little or no attempt was made by the original founders of EF! (UK) to regulate or determine the philosophy or actions of new groups. Garland outlined the range of actions of his local Mid-Somerset group during the summer of 1991:

> We did a mahogany action...went to B and Q and we worked with FoE there. I tried to get on the roof and, um, got pulled down....The most successful thing we did was the health food boycott sheet which went international....We were setting up a network for animal rights' groups in EF! That was an important aspect also.
>
> (Garland in interview)

South Downs EF!, one of the most active groups between 1992 and 1998, not only carried out the first 'Carmaggedon' action, disrupting a local road opening, but targeted British Telecom. *Green Line* reported:

> 'Recycle the Yellow Book Wall' demanded South Downs Earth First!...when they built a metre-high wall of telephone books along an eight-metre stretch outside British Telecom House, Brighton.
>
> 'BT profits average £95 per second,' exclaimed demonstrator Alec Smart, 'yet incredibly, they take no initiative to recycle their expired phone books.' The directories are difficult to recycle because of the latex glue used in the binding....Earth First had initially planned to stack the thousands of directories they had collected into various phone booths about town, but fearing unfavourable headlines (e.g. 'Reckless Greens stop desperate mum from dialling 999') they lined them up in front of the BT offices instead. The BT security guard, described as 'apoplectic' by onlookers, told the demonstrators: 'You can't leave those here!' But they did.
>
> (1992 (95): 4)

Local activity has remained varied and local group structures minimal. Only a minority of activists owed a 'primary loyalty' to EF! and local 'groups' can be seen as networks of networks, acting as contact points into the denser local green circles. In the same way that EF! (UK) emerged as a national body, however biodegradable, by using green networks, local groups used local equivalents to mobilise activists for direct action. Thus Begg noted: 'We'd phoned round our huge list of contacts. We let the hunt-sabs know....Leeds Cycling Action Group are usually good for people on anti-roads actions....Third World First have helped us out a lot as well.'

Typically, 'local' EF! (UK) actions which I participated in, between 1991 and 1994, might involve 50–100 activists. For example, early attempts to block a quarry at Whatley in Somerset drew in local campaigners, Green Party activists and those such as Garland who identified themselves as 'EF!ers'. By the late 1990s, however, thousands of activists were more likely.

Earth First! and the anti-roads movement: 1992–5

In 1992 EF! (UK)'s issue-focus shifted to anti-road campaigns. While local groups campaigned on diverse issues, activists increasingly mobilised against car use in general and, increasingly, road construction in particular.

During the summer of 1991, Torrance, moving from a south-coast town to a busy part of London, seems to have been struck by a personal antipathy to motor vehicles:

> I was really moving away from forest stuff, getting more into the whole idea, living in London, of thinking: 'Fucking hell, what is going on in this city with all this traffic going around me? Something has fucking got to be done about it – if you really want to start things going – outside everybody's fucking doorstep all over the country'....Most people own a car, and I know from speaking to people, cars and transport is a very sensitive issue, so I thought this is one that has to be worked on.
>
> (Torrance in interview)

In articulating this concern, he was able to tap into a growing anti-car sentiment among greens (Torrance 1991). Both radicals on the fringe of the green movement and environmental pressure groups were increasingly critical of the environmental and social costs of increasing road traffic (Control 1991; Finch 1991; Greenpeace International 1991). In August 1991, Torrance organised a campaign meeting to plan anti-car actions with

> key people in the then very thin anti-roads/traffic/car movement, people from ALARM, some EF!ers from EF! groups – I think some one came from South Downs – Davy came up, a women came up from the newly formed Littlehampton Earth First!...Karen Noble – and she has an intense hatred of the motor vehicle to a supreme level, which I was very impressed with – and

> Angie came, and we had a really, just a really amazing brainstorm on setting up a new roads campaign.
>
> (Torrance in interview)

Laughton, noting resource factors and the importance of personal links, observed: 'I think the main reason they got into roads was because of a very strong woman called Karen Noble, who was very anti-roads, has money, is quite attractive.' The meeting focused on the need to create 'a grassroots resistance anti-car culture' and inspired a number of initial small-scale iconic actions (Torrance in interview).

Torrance laid out the case against the car in an October 1991 article for *Green Line*. Stating that 'Earth First! will launch a "ban the car" campaign entitled Carmageddon', his critique was broad and passionate. The greenhouse effect was identified as a consequence of the 'great car economy':

> The car's effect on our environment doesn't stop at direct exhaust pipe pollution but also extends to extraction of minerals for production, oil use including transportation dangers, and disposal of old cars...each year 23 million tyres are discarded in the UK alone and 28 million gallons of motor oil go missing, presumably finding their way into our freshwater systems.

The Gulf War, the ill-effects of advertising and reductions in biodiversity were noted in this article, as also was the idea that the cultural politics of car ownership disguises violence:

> In today's society the car has become the norm, a symbol of wealth and prosperity. In reality it is the most destructive item an individual can possess. The worldwide slaughter of humans and other species amounts to hundreds of thousands each year, with both the injury and death figures on a scale equivalent to any war.

Torrance expressed the conviction that green criticism of the car had been muted by industry pressure, noting:

> In the general level of apathy and hypocrisy of most green campaigns to date, the idea 'responsible car driver' has been cooked up. The car industry has launched the 'green car', which emerged like the phoenix from the ashes, to all but the most realistic of people.
>
> (Torrance 1991)

The Lea Valley (London) EF! group tried unsuccessfully to disrupt the official opening of the Dartford road-bridge, using Sea Action boats. Unfortunately, the police speed boats were rather faster than the borrowed peace-movement vessels. On the last Saturday before Christmas 1991, South Downs EF! 'carried out the first road blockade in the Carmageddon campaign' in Brighton, suffering one arrest (*AU* 1991 (2): 1). Shane Collins, a Green Party activist and founder of Brixton EF!, described the first London blockade on 15 May 1992:

> We invented this Carmageddon day and printed loads of flyers, sort of saying: 'Are you pissed off with cars, do they get up your nose?', and I guess about 70 or 80 people turned up at Victoria Embankment. We went up to Waterloo Bridge and sat down a few times....[There was a] constant police [presence], they were there and knew we were coming, and they tried to shepherd us onto the pavement and push us off the Bridge, and we sort of radically sat down in the road.
>
> (Collins in interview)

On 1 August 1992 in Hyde Park, a Mini was ritually smashed, and extracts from Heathcote Williams' book *Autogeddon* (1991) were read out (Collins interview). Williams' text has been described and criticised as a '[c]offee table anti car book. Interesting quotes section but the poetry's dreadful. The gratuitous Holocaust metaphor is particularly crass and offensive. People were driven into the ovens, they didn't drive there of their own volition' (Control 1991: 10). The initial anti-car meeting led to the creation in April 1992 of EF! (UK)'s Reclaim the Streets (RTS) campaign, which aimed to 'do imaginative and non-violent direct actions, and to reclaim the streets of London from cars and traffic and give it back to people' (Anon. 1992: 4). RTS, despite a break in activity so as to concentrate on specific anti-road building campaigns, has gone from strength to strength. The tactic of the street party, where pedestrians occupy and enjoy road space, gradually evolved from these early carmageddon actions, though a version had also been carried out by Commitment in the 1970s. Participants were mobilised by the promise of 'fun', and mass participation, in turn, lowers the cost of involvement by making arrests less likely and police response more difficult to enact with efficiency. Each action led to a more ambitious objective. During the largest to date, in July 1996 7,000 protesters occupied a London motorway (Bellos and Vidal 1996: 7; Honigsbaum 1996: 28; participant observation). During the summer of 1998 simultaneous parties were held in Brixton and Tottenham, each drawing around 3,000 protesters.

None of the actions over 1991–2, with the possible exception of the Dartford road-bridge protest, specifically opposed new roads. Yet the general green condemnation of the motor vehicle coincided with an upsurge in local – usually single-issue – anti-road campaigns. Such campaigns, as already noted, predated EF! (UK)'s mobilisation, but at that stage rejected direct action for legal and symbolic campaigning. In contrast EF! (UK) explicitly appealed for direct action in campaigns against road building and sought to link road construction to global environmental problems. EF! (UK)'s *AU* identified three road projects in which the EC Environment Commissioner had intervened against the plans of the Department of Transport (DoT). An appeal was made for support in 'confronting the bulldozers' if construction went ahead at the east London River Crossing, the M11 in east London and the planned extension of the M3 at Twyford Down (*AU* 1991 (2): 3). The Twyford campaign, to which I turn in the next chapter, was to have a dramatic effect on EF! (UK) and effectively launched the anti-roads movement of the 1990s. While culturally and geographically distant, Hampshire Down land and the New Mexican desert are linked by EF!'s story. Ironically, EF! was to have a far greater impact in the rainy south of England than in the arid region of its genesis.

4 Twyford, other tribes and other trials

> A great saying runs: 'Mankind marches to annihilation under the banner of realism' – we must resist the weasel words of 'realism' at all costs – after all, it was a 'realistic' attitude – as Andrew Lees admitted in the *Guardian* – that led FoE to abandon Twyford Down, and that leads people into passivity and defeatism on nearly every occasion. Some pride in our achievement is warranted here – we have given many people in the UK – and especially within the environmental movement – a concrete illustration that direct action works and produces results. This is an antidote to the prevailing attitude of powerlessness and hopelessness that keeps people down and the planet under attack.
>
> (Letter in *Do or Die* 1994 (5): 94)

Twyford

Twyford was the launching ground for the direct action road protests of the 1990s and had been a high point of 1970s anti-road campaigning, thus being instrumental in delegitimising road-building programmes in both decades. Campaigns involving hundreds of activists initially, and thousands later, using increasingly dramatic tactics stemmed from the Twyford experience. 1993 saw the start of protests in Glasgow, Jesmond Dene in Newcastle, the M11 in east London and Solsbury Hill, near Bath, in the West Country. In 1994 the M65 in Lancashire was the target of action. 1995 and 1996 saw the protest move to the Newbury Bypass in Berkshire. 1997 saw dramatic evictions from the A30 route in Devon, where the last tunnellers to be removed were feted by the media. Dozens of smaller protests have been undertaken throughout the 1990s.

St Catherine's or Twyford Down in Hampshire, near the medieval city of Winchester, is an area of chalk down land, teeming with wildlife and archaeological monuments, including an Iron Age hill fort (Bryant 1996). Local campaigners disrupted a 1970s public inquiry (see pp. 32–3), frustrating plans for the M3 link for some years. Opponents of the M3 during the 1970s disagreed strongly over where the motorway should go, and by the 1990s some defenders of Winchester city-centre, and the local public school Winchester College, were happy to see

Twyford Down as a substitute route. Barbara Bryant, a local Conservative councillor, has chronicled attempts by the Twyford Down Association (TDA), using every legal means available, to prevent construction (Bryant 1996). Despite the intervention of EC Environment Commissioner Ripa di Meana, such orthodox tactics failed.

Direct action against construction began shortly before the 1992 General Election:

> [A] shocking but true story about Twyford, yeah....Earth First! and Friends of the Earth got involved with it and got involved with it for one reason only. They thought they would have an easy victory, they thought there would be a 1992 Labour election victory and Labour was going to cancel it. The point I would say to Earth First!'s credit [is that] they stuck with it when that didn't happen. And Friends of the Earth didn't, right?
>
> ('Mix' in interview)

Davy Garland observed:

> Winter of 1991, December, Twyford Down started coming into the view....Contact was made with Dave Croker of the TDA, and we had the opportunity of us getting involved in Twyford along with Friends of the Earth, so went up to the Cooltan, up in Brixton, and I basically had a meeting up there....It was mostly Earth First!ers and David Croker. He turned up in his little sporty car and we were talking about making a camp [and] defences. David Croker was offering a kind of old army base....The rail bridges were coming up. So we were going to have that as a major action.
>
> (Garland interview)

Although other members, such as Bryant, were hostile, David Croker, a former Conservative councillor and leading TDA activist, who (see p. 33) had been a leader of the 1970s disruption, made vigorous attempts during 1991 to build a network of support for direct action. On 18 February 1992, EF!ers along with TDA supporters carried out the first direct action on the planned route:

> Six protesters from the radical green group Earth First were arrested during a weekend of protests against plans to extend the M3 motorway.... Environmentalists and local people had occupied two Victorian bridges over the Waterloo to Weymouth line which had been due for demolition.
>
> (*Independent* 19 February 1992: 1)

According to Garland, TDA members managed to distract the police while EF!ers made for the bridge.

Eventually the first permanent anti-road protest camp was established at Twyford. The camp's tactic, mentioned by Garland and independently suggested by other individuals, seems to have been inspired by peace campaigners, most notably the Greenham women and the Australian rainforest movement (Begg 1992: 9–10; Roseneil 1995: 172). Torrance noted:

> We talked to David Croker about setting up a camp, again drawing inspiration from elsewhere, drawing from the bush in Australia and US and peace camps here, and I then thought, 'Wow, just imagine, one day, fucking hell we could have all these camps over here against roads…'. And there you go, it happened.…So I mentioned the idea of a camp to Roger [Higman, the FoE transport campaigns' officer], and he said, 'No, no. Silly idea. You know the last thing we want is loads of people trampling on a SSSI' [site of special scientific interest]. So, anyway, he won me around to that on some point. But I was kind of still looking – but not as seriously as I should have been – at camps in other places.
>
> (Torrance in interview)

Begg noted:

> I had written an article for *Green Line* in which I had said: Peace camps, what a brilliant idea.…Why don't we have camps at other places – factories, construction sites, animal-testing labs? And I thought this was a great idea, but being as I was, like, somebody living in a house, I wasn't going to do anything about it myself. But when I heard that somebody was setting up an anti-road camp at Twyford, I was very excited.
>
> (Begg in interview)

Ironically, despite Higman's (reported) comments, the very first anti-road camp was established by FoE:

> FoE decided on a symbolic occupation of the threatened Water Meadows.…[It was] launched with Merrick, Jonathon Porritt, and Andrew Lees of FoE being chained across the site. The weather instantly turned cold and wet, and local people fell into the habit of taking hot coffee and breakfast down to the site.
>
> (Bryant 1996: 189)

FoE received an injunction from the DoT and abandoned their protest for fear of being forced to pay huge legal costs (Freeman and Secrett interviews). According to Sheila Freeman, who worked for FoE and later became involved in EF! (UK) and the M11 campaign:

> [The injunction] was served on the poor women from the finance department! So that was the big mistake that FoE made, and always regretted it, sure: 'We can't possibly afford this injunction. It will be £1,000s and £1,000s.' So that was it – out of Twyford, there are plenty of other places to think about. Instead of explaining to people why they did it, they just dropped out of it. There were lots of people, a year later, who had no idea why they left, and even the ones who did know thought it was done wrongly.
>
> (Freeman in interview)

Most EF!ers were damning in their disagreement with the FoE's decision; but Becky Lush, who was active in EF! and a Donga (see below), noted:

> A lot went on that it has been convenient for the direct action movement and Earth First! to hype, and, just as FoE are accused of trying to get members on their side, so are EF! It is not financial or anything like that, but it is convenient for people to press direct action. Direct action: 'We're the best, FoE copped off and fucked off…'. All that happened was Porritt turned up and chained himself to a bit of fencing, and then they got an injunction and fucked off….I found them really nice. Really different political views, obviously. They don't really know what bastards the police are yet. But I didn't, until I got involved, and they think changes in the laws save everything. They have got weird political analysis. [Yet] when you look at it, they really tried to help the TDA. They set up this bizarre 'we are the middle-class, we are representative of middle England' and extremely media-obsessed [camp]. They found it uncomfortable, but they were fucking there. They used symbolism very powerfully, they used the media skilfully and stopped the works with their own tactics….After three days the police cut this very-easy-to-cut chain and they got slapped with this injunction [meaning that] they could be fined as a company and have their assets taken. I can be gung-ho and say I will go to prison for as long as it takes. That is my individual decision, [but] as a company they have to make decisions….They campaign on everything…unless they wanted to cease to be FoE and throw all their money into this.
>
> (Lush in interview)

In the summer of 1992 the camp was re-established:

> The camp, at the time, was very much made up of groups of travellers who had [just] ended up there….First time I went down there, there were only half-a-dozen people – literally a couple of travellers who had been involved in other stuff and looked a lot older than everybody else; a woman with a

> couple of kids, on the run from social services; an ex-army drop-out; also people coming up from EF! in Brighton.
>
> (Anon. in interview)

Thus as well as attracting green activists Twyford and later camps provided a magnet for new age travellers and for individuals simply seeking to escape from authority, family commitment or urban unemployment. Becky Lush visited in 1992, and her description captures something of the spirit of the place:

> [I] turned up at Twyford Down, parked at the Dongas and walked up. The first thing we saw was this traveller's van. We poked our heads in and there is a guy watching fucking *Neighbours* on a miniature TV, and I thought 'What's going on...? The landscape was very weird: I had never seen anything like it, this bizarre set of gullies, and I thought, "Wow! This is good"...'. [I] got to the top and thought, 'This is even weirder.' There were teepees, and I had never seen a teepee before. It was, like, people lived in teepees, and there were all these very charismatic, bolshy, Donga types.
>
> (Lush in interview)

Later she joined the camp, noting:

> I went in this teepee, and there was this grumpy old man. It didn't seem particularly welcoming or friendly, but it was interesting and the place was stunning, but nobody talked about the road whatsoever. Nobody talked about what was going on. I had no inkling of the campaign there whatsoever. To me it was travellers living on a hillside.
>
> (Lush in interview)

A camper called 'Jazz' claims that the Donga identity has been reified into a badge of identity that fails to show the loose affiliations of the original campers:

> Initially people at Twyford were quite a muddled bunch. There were pagans and there were anarchists, and there were people coming from all walks of life and from the established green movement, from FoE and the Green Party. So I initially came across something called EF! at Twyford, but it was so muddled in that I didn't really pay an awful lot of attention to it. The identity was the Dongas'...ancient medieval trackway.... We had a meeting and passed a little totem pole around. We declared it an autonomous territory and called ourselves – loosely, all this collection of different people – 'the Dongas Tribe', which has progressed into something quite silly, but at the time was meaningful....Unfortunately, I do feel that that identity has become something quite different and it has become a tool to disempower

> people, simply because there are groups of people saying 'I am a Donga' and placing themselves on some sort of pedestal....At the beginning anyone who came to Twyford Down and did anything, you know, was a Donga....I think, now, it has been recuperated into a fashion statement/identity/ideology.
>
> ('Jazz' in interview)

In turn Alex Plows, another camper at Twyford, argued that an alchemy was achieved:

> This was an extraordinary meeting of strong people and the inevitable consequence of this, apart from catalysing an amazing movement, was the odd power clash. It's out of these personality clashes that this whole 'the Dongas were like this' or 'the EF! lot were like that' comes from. Identities simply weren't that distinct, never were, still aren't, shouldn't be. Belatedly creating, and then polarising, two groups into two views is not fair; it hides the truth, that all of us together were doing some incredible, original, brave stuff, rattling through some vital learning curves at an incredible rate. Yeah, initially some people were more into living on the land, others were more into days of action, but quickly everyone learned that it took all of it together.
>
> The identity was the Dongas as trackways not tribe, and anyone who felt for the land was part of this group identity...that counts in everybody who was ever there and fought for the land. Chris Gilham and the rest of the TDA are as much a part of this as anyone else. All of us who were there have a spiritual and physical connection with the place and through the place to each other.
>
> Personal communication

The linkages, sometimes conflictual, sometimes co-operative, between EF! (UK) and other radical green activists – local campaigners, environmental pressure groups, travellers and tribalists – were to be a constant feature of anti-road protests in the 1990s:

> EF! got told to leave by the Dongas tribe. Well, a few people [did], because they were upsetting the karma of the place. This was a couple of days before Yellow Wednesday. [DW: So Dongas are distinct from EF!?] No! It became so around that split because of dominant people in the Dongas. Those sympathetic were in the background....EF! brought people to Twyford, that was what really did it....They were the network who did it. Other networks wouldn't touch it. FoE wouldn't touch it, none of them would touch it, and that is still the kicking power of EF! It can produce people who are willing to do something.
>
> (Anon. in interview)

Although some Dongas, such as Graeme Lewis and Becky Lush, identified themselves as EF! (UK) activists, there was friction between the two bodies. Direct action, in particular, led to conflict because it brought reprisals from the security guards:

> There was a point when loads of EF!ers came down and trashed lots of fencing. . . . The security guard came down and burnt down this watchtower, and that was really scary. That was 12 o'clock, and we said: 'Fuck it. We aren't going to go to bed.' So we made lots of porridge. Then the next day. . . . barricades around the Dongas were set alight with diesel. They were green, so they didn't really catch, but it was very fucking scary. . . . It would have gone *whoosh*! and set tents and the marquee, where lots of people were sleeping, alight. And after that the Dongas said there were to be no more direct actions. . . . There was some kind of meeting, and the strongest characters made it clear they did not like direct action happening. . . . It was antagonising the workers, and you thought, 'Fair enough'. And then you thought: 'Why are we sitting here, in this lovely little camp, if we are not going to stop the road?'
>
> (Lush in interview)

Some Dongas were neo-tribalists with a complex mythological commitment to the land, believing that Twyford Down was the site of King Arthur's Camelot. Such heterogeneous beliefs, combining diverse pagan myths, came from a wider travellers community and incorporated a rich sense of ritual practice (Lowe and Shaw 1993: 112–24).

A compound was built by contractors and direct action continued through the autumn:

> I did my first direct action armed with tambourine and drums, ran down the hill. . . stopped and watched these nutters run out in front of wheels taller than myself. . . . I thought: 'What are they doing? I can't do this, it is terrifying. I will get run over.' The next day I stood in front of a digger. I smiled at the driver and he stopped his machine. . . . Being a woman is easier in direct action because they usually don't like to intimidate women. So I found it quite easy – just shared fags with them.
>
> (Lush in interview)

On 'Yellow Wednesday' in December 1992, the camp's population was dramatically and violently evicted:

> I was woken up by Busker Paul saying: 'Fucking hell, there are 100 people in yellow jackets marching around. I think they have come to do some

> surveying or something.' We hadn't seen security guards before, it was all totally new.... The Dongas were picking up mud and just throwing it in the faces of security guards. That really wound them up, and wasn't a particularly clever thing to do. But it was obvious the big day had come, and people were really, really upset, and then I saw one guy jump the security lines.... You would climb in front of a machine and they would catch you by your hair – I had long hair – or a finger or your wrist, and they would drag you through hawthorn. You would be scratched to fuck, and then they would drag you over the flints.
>
> (Lush in interview)

The camp, at this time, numbered just twenty individuals, but more arrived for a planned party and the protest continued until the next evening.

> It was surreal. Suddenly the Dongas [camp] was huge arc lights, artificial lights.... We had a wild party. The moon started to disappear, and we managed to get out some sloe gin that I had been making. Lunar eclipse. It was quite weird. The Tofu Love Frogs played, and they said: 'Hey everyone, let's go and say hello to all the security guards.' Everybody went over with fiddles...cutting the razor wire down and trampling it. It was amazing, really good, and we started dancing under their noses.
>
> (Lush in interview)

After the eviction, most of the self-proclaimed Dongas moved to a woodland 'about fifteen miles from Winchester' for the remainder of the winter, before travelling to the west of England (Lush interview). The camp was re-established by six EF!ers the following February (1993) in Plague Pits Valley (Lush, Anon. and 'Jazz' interviews). After both the Twyford Down Association and the Dongas had retreated, Lush and Emma Must, a librarian, helped set up a new group – the Friends of Twyford Down – to oppose construction of the road:

> The TDA, after the EC gave their decision, said: 'Right, we have done our stint, we have fought it through all the legal channels.' But there were a handful of TDA members who thought that this is not it, [who] thought they could stop it through the direct action stage. The Friends of Twyford Down was set up to continue the fight.
>
> (Lush in interview)

Anti-road actions occurred on almost a daily basis throughout 1993. DoT affidavits reproduced in *Do or Die!* suggested numerous acts of ecotage: 3 March saw 'Dumper truck locked onto on Twyford B-roads, stopping work in cutting and water meadows for 1 and half hours'; on 4 March, fifteen people from 'Camelot' EF! and Friends of Twyford Down 'entered cutting at 7.30 a.m., immobilised machinery'. On 5 March, '200 people gathered for a weekend of action and inspiration co-ordinated by *The Ecologist*...much drumming and fun'. On Saturday 5 March action intensified, the weekend boosting the number of activists. On the Sunday a 400-strong rally in the motorway cutting was addressed by green movement speakers and the Liberal Democrats' environment spokesperson; chalk which the contractors had dumped in the water meadows was then returned to the Down via a human chain. 'Police looked on in horror, seeing respectable people ripping down barbed wire.' Monday 8 March saw sixty people entering the cutting to lock onto machinery, halting work yet again (*Do or Die!* 1993 (2): 10–11).

Daily protest continued throughout the spring and summer, costing contractors an estimated £2 million. In July Tarmac brought an injunction against seventy-six named activists, threatening them with imprisonment if they entered the work area and making them liable for the claimed £2 million. This was widely thought to have been aimed at local, often middle-aged, protesters who could be made bankrupt and lose their homes. Two days after the injunction 600 activists, including a group of 'injunctees', took part in a mass trespass on the site. 'This was the biggest Twyford action yet and it seemed (happily) more like a festie than anything else, [with] a healthy mix, sabs, travellers, doctors and EF!ers' (*Do or Die!* 1993 (3): 23). On 23 July seven injunctees, including Torrance and Lush, were sentenced to a month in prison, creating a considerable media stir. The seven were visited by Carlo Ripa di Meana, the former EC Environment Commissioner, to the considerable embarrassment of the UK government. They were released after just thirteen days.

Protest, albeit more sporadic, continued into 1994. The fight for Twyford Down made use of every legal and extra-legal tactic, from appeals to the EC to the disruption of a public inquiry to 'digger-diving' and the incineration of heavy earth-moving equipment. Benny Rothman spoke at one rally to encourage militancy, yet even the appearance of this grandfather of anti-capitalist green action from the 1930s could not preserve the hill.

While Twyford was the first arena for direct action anti-road protest in the 1990s, EF! (UK) activists were instrumental in creating permanent anti-road camps and maintaining campaigns at over a dozen other sites. An EF! group 'Green Man' helped initiate action at the M11 in east London (Mercer 1994: 118). Five members of Bath EF! started a camp on the site of the Batheaston Bypass ('Clare' in interview). Torrance noted how 'very often Earth First! comes together around an action or an issue, like up in Wymondham in East Anglia and [an] Earth First! group really grew and blossomed around the Wymondham Bypass'. By 1993 EF! (UK) was absorbed by the wave of anti-road protest that it had helped to launch.

M11

The next large-scale campaign took place in east London. The emphasis, as in the early 1970s, was on 'homes before roads', and buildings were squatted in an attempt to stop them being bulldozed (Anon. 1996b; Jordan 1996). While 'nature' in the form of the famous chestnut tree (see pp. 75–6) was, on occasions, directly defended, the social ill-effects of motorways and the menace of car-culture were the significant factors motivating action. The link-road cut through two distinct parts of the city: Wanstead was a quiet middle-class commuter suburb; in contrast Leyton and Leytonstone contained high-density working-class housing apparently, according to one activist, 'badly neglected ever since the proposal for the link-road first blighted the area' in the 1950s. 'To look at it now, you could be forgiven for thinking the best thing for Leyton and Leytonstone would be to demolish them and grow a forest' (*Do or Die!* 1996 (4): 21). The involvement of EF! (UK) in the M11's urban environment shows just how far the network had moved from EF! (US)'s original obsessional interest in wilderness. Even so, Roger Geffen, who helped launch the protest, noted: 'M11 was the Cinderella – it didn't have an area of beauty, it didn't have an 8,000-year-old wood, it was a run-down working-class area and nobody thought it [had] much hope as a campaign.' FoE had deprioritised the M11 partly because of its lack of nationally important environmental sites. Secrett, who was then FoE Director, commented: 'You can't do everything or you stretch yourself too thinly. The M11 was not one of our priority road campaigns because we had not had a long-term community involvement' (Secrett in interview).

Local groups had long campaigned against construction. Residents, often old-age pensioners whose homes were to be demolished, wrote tear-filled tracts, presented as evidence against compulsory purchase orders in the 1989–90 public inquiry. Louise Swift, one such resident, observed typically:

> Although I cannot go out I am never lonely, for in this road [there] is a lovely spirit of community. If I am in my garden everybody stops for a word or two, Asian, West Indian, Greek, Turkish, African and English....I ask why must they pull the houses down for one more roadway and break up a lovely community like ours, and cause old people to fear the future?
>
> (Quoted in Field and Drury 1995: 13)

During the last stage of the final public inquiry the inspector, Colonel M. F. Davies, a retired soldier, had hecklers ejected by the police. Their 'non-violent resistance to this eviction by force was a small preview of what was to come' (Field and Drury 1995: 13). By 1993 the local campaign had stagnated:

> There had been no public action....They had been through public inquiries, they had been a small bunch of mainly elderly men. I remember Paul describing their meetings as being like you were one of the cast of *One Flew Over the Cuckoo's Nest*. [They were] elderly eccentrics who obviously had not made any links into attracting community support out there.
>
> (Geffen in interview)

Campaigners at the north end of the route, where construction was due to begin, seemed to have largely given up the fight by June 1993:

> I was encouraged when Paul Morotzo got involved, coming down from Twyford and...was quite keen to make something happen up there, and he dragged me up there....We met Patsy Braga. She was living in one of the houses that became Wanstonia. She was pretty much a lone campaigner at the Wanstead [north] end of the route; the other people in the 'No M11 Link Road' campaign were at the Leytonstone [south] end.
>
> (Geffen in interview)

EF! (UK) activists and others organised a march along the proposed route in mid-August 1993. Squatting had begun earlier.

> Steve Johnston who I knew through the LCC [London Cycling Campaign]...had been to Twyford and he had set up a bender in the garden, I think, of Henry Cox who was trying to encourage as many squatters to move into the area as possible. He [Johnson] called a first action up at the Green Man roundabout and did a peaceful occupation, and a few EF!ers from Twyford went up.
>
> (Geffen in interview)

Work was due to begin on 13 September when around 100 protesters occupied a vacant house which was subject to a DoT compulsory purchase order: 'By the end of the day, [it had] had its roof repaired, and was covered in banners. 1–0 to the campaigners' (*Do or Die!* 1996 (4): 22). Tree-sitting, 'digger-diving' and other forms of delay met the contractors Norwest Holt, whose employees took several weeks to clear a small woodland. Activists from across the UK flowed in, though local support was still limited.

> What changed all that was the sweet-chestnut tree and the importance of Jean the lollipop lady...telling school children their tree was being cut down, and we had all these school kids coming and lying down. The rest and all of their parents got stuck in. It took them a week to get the fences up because of the school kids, and then everybody came and pulled them down

> again. That was the point when we suddenly discovered lots of community support.
>
> (Geffen in interview)

Local people believed that the M11 would tunnel underneath George Green, and the tree, said to have been 250 years old, would be saved. The small cut-and-cover tunnel, in fact, threatened its swift demise. On 2 November 1993 the contractors built a fence around the tree and prepared to bring it down. Four days later local people tore the fence down. A tree-house was built in the chestnut and letters of support addressed to its occupants gave it the status of a legal dwelling. This slowed efforts to shift the protesters and created an impact in the media, accelerating the scale of action:

> I think we may have made legal history: the first tree-house to be declared a dwelling in British law. . . . That started to run as a little story in its own right, and also bought us time because the Department of Transport [was] completely snagged up with the fact that they hadn't gone through the proper process of transferring occupancy of the land adjoining George Green.
>
> (Geffen in interview)

The DoT eventually gained a court order to oust the tree's occupants, and on 7 December, early in the morning, 400 police arrived. A high-hoist hydraulic platform or 'cherry-picker' was brought in at 11 a.m.

> It may seem strange that it took the contractors so long to get a cherry-picker to the tree, however this can be explained quite simply. The contractors fearing that monkeywrenchers would be in the area put their cherry-picker under 24 hour guard, and being completely paranoid hired another and hid it down the road in a school car park...the hydraulic platform started to fall apart when they attempted to move it. Without hesitation they drove down to their 'secret' backup platform only to find it wrenched! Cherry-pickers are rather sparse in east London, especially for companies who have a reputation for 'not looking after hired machinery'.
>
> (*Do or Die!* 1994 (4): 25)

Despite the loss of the tree, after a nine-hour police action costing around £100,000, campaigners remained buoyant. Media attention mushroomed, with pictures of the tree appearing on several dailies' front page. A row of houses next to George Green was squatted and the campaign continued.

> Patsy Braga lived in a large late Victorian House, next to George Green, it was one of a row of houses that hadn't been demolished. Those three build-

> ings plus two burnt out shells were Wanstonia and by the end of December Patsy was served notice that she had to get out. She said, 'All right. It is time I became a squatter in my own house,' and she invited everybody else in as well. And she rather manipulatively said to the media: 'I just moved out. I found that my house had been squatted [and] they seemed like a nice bunch of people, so I just moved back in with them.'
>
> (Geffen in interview)

Mimicking the Ealing film *Passport to Pimlico* and 1970s' tactics, symbolic independence was declared, bringing the Republic of Wanstonia into being.

> The independent free area worked well in the press....[We were] writing to Douglas Hurd that we [Wanstonia residents] were dropping out of the UK and writing to the UN asking for recognition, and producing passports, and fun-and-games like that, announcing it at the High Court [and] getting Simon Fairlie to write something in bastardised prose that was a version of the American Declaration of Independence. But we twigged it enough to get lots of environmental messages and anti-government messages [publicised], and it was great fun.
>
> (Geffen in interview)

Defences were modest but contributed to the accelerating acquisition of skills, culminating in the east Devon tunnels which attracted so much media attention:

> A lot of barricading was going on...most of it was not terribly sophisticated, people were digging trenches somewhat naively...we were completely inventing it without any kind of plan. A few people had sussed ideas for creating concrete lock-ons,...it was all fun, community spirit even if it didn't make a lot of difference on the day.
>
> (Geffen in interview)

The 'No M11' campaign launched 'Operation Roadblock' – a month of daily direct action, which started on 15 March 1994. The date represented the 'Ides of March', when Julius Caesar, the leader of a great road-building empire, was assassinated. The aims, as at Twyford, were to wear down the contractors with continuous action and to recruit new protesters. Operation Roadblock was only partially successful and there was a certain amount of 'burnout', with seasoned campaigners becoming exhausted. It built on 'Mix's proposals for sustained well-organised action, bringing in grassroots' support to replace increasingly pressurised core organisers. Much of the campaign, though, jarred with 'Mix' and other militants because it stressed symbolic action and strict non-violence.

Operation Roadblock also attracted to the M11 seasoned green activists from London and beyond.

> We hadn't made much link with the cultural and festivals side. It was mainly green-movement people....There were certain local London FoE contacts who encouraged the rest of the group to turn up...if a FoE group turned up in force, even if it was only five or ten of them, that could help numbers brilliantly. EF! groups would tend to arrive either as ones or twos or as whole groups – in the case of Lancaster or Leeds – and stay several days. The rest you didn't get to know who they were. A lot of them would just tend to stay for one day, but they were probably people who were already active in some green, social-justice or peace campaign.
>
> (Geffen in interview)

April saw the end of Operation Roadblock, and action moved to the southern end of the route and Claremont Road. A bold outrage against McGregor, the Transport Minister, was used to publicise the campaign.

> It was a fucking beautiful day, everybody kipped out on Paul Morotzo's houses, forty people in a room not much bigger than this, we took this ladder through Highgate Woods and got there at 6.00 in the morning. It was beautiful, it was very satisfying. The idea was so spot-on, a road through his house with love from Claremont Road, because we wanted to get Claremont Road on the map.
>
> (Geffen in interview)

Protesters climbed onto McGregor's roof in Muswell Hill Road and unravelled a life-sized imitation motorway, to the delight of the press. Glen Canyon Dam had once again been symbolically breached.

Claremont Road was occupied and heavily defended. It was a riot of colour, packed with murals, sculptures made from distorted waste, a giant chessboard, a spider's web of nets stretched across from the roofs and a 100-foot high scaffold tower. The nets had been gathered from circuses and it demanded some courage to climb across them (Hunt 1995: 128). The symbolic and the strategic could not be separated: the scaffold poles were greased, presumably with an appropriately vegan medium. When the street was evicted in strikingly paramilitary style, the last campaigner, Phil McLeish, managed to avoid capture for several days. It was simply too dangerous to attempt to remove him from the tower. Art became a weapon and music brought in party-goers. Heather Hunt, a former Greenham activist who has chronicled her experiences at Claremont, observed: 'Every Sunday night, there was a street party, and musicians came and folk-singers came, and there was dancing [in] the street.' (1995: 125).

This final eviction did not occur until November 1994, many months after the assault on McGregor, and resistance continued at other squats into 1995. As had been the case at Twyford, massive costs were incurred by the authorities; thousands of individuals took part in direct action, and lives were transformed as a result.

> We cost them £6 million, we had the longest eviction...well, all right, the longest greenie eviction in bloody history...at least equal to the one up at Molesworth in 1985 [when a peace camp was removed by the army]. Yeah, I mean there were the same number of cops, etc., etc., as there were Royal Engineers brought into Molesworth.
>
> ('Mix' in interview)

After much delay, construction went ahead. The Romans conquered, but Caesar was further wounded.

Solsbury Hill

Solsbury Hill is on the very spur of the limestone Cotswolds and the outskirts of the city of Bath, where the Romans had established a temple to the deity of the hot springs. Solsbury is close to the Roman Foss Way, a first-century 'motorway' and military border. Like Twyford Down, it is topped by an Iron Age hill fort. Peter Gabriel, a rock star of the 1970s and 1980s, who lived nearby, immortalised the territory in his song *Solsbury Hill*. Thus Solsbury Hill is saturated in powerful symbols of the British countryside, history and prehistory. In 1994 a campaign was launched to prevent the construction of a road across the lower slopes of the hill.

Bath EF! and 'Save Our Solsbury' (SOS) argued that the road would damage a designated 'area of outstanding natural beauty' and 'destroy numerous badger sets, newt ponds and pass through Little Solsbury Hill', as well as turning 'one of the most precious water meadow eco-systems in the country into barren wastes of mud and rubble' (*Do or Die!* 1994 (4): 1; see also *Action Update*, 'Dijon to Belfast Euroroute', 1994). Instrumental in launching and sustaining direct action was a teenage EF! (UK) and Green Party activist named Tania de ste Croix. Not only was she a highly active campaigner, probably the most prominent Green Party member in the city during the early 1990s, but her own home was eventually demolished to make way for the road.

The green movement has been strong in the area since the late 1960s. Nearby Stroud had elected more Green Party councillors than any other area in the UK during the 1980s and had been the scene of a successful tree-sitting when a supermarket road threatened local trees. Some of the best results for the Party during the 1989 European elections were achieved in nearby constituencies. Both local amenity campaigns and direct action networks were strong when the Solsbury Hill campaign started. The Bristol Environmental Action Network, affiliated to EF! (UK), contained activists who had fought the construction of a Bristol supermarket at Golden Hill, using a permanent camp and tree-squatting tactics. Bath EF! also had been established in the early 1990s. In contrast to Twyford and the M11, a 'counter-movement' in the form of a local campaign group, based in the village of Batheaston, supported this road, arguing that it would act as a bypass taking traffic from their busy high street.

Bath EF! and SOS believed that the road was part of a EC road project which would bring large traffic flows to the area:

> The 'bypasses' are actually sections of a euroroute which, until recently, had been kept secret from Bath locals and the public. Much like the Watford bypass which quickly materialised into the M1 these bypasses are simply a motorway being built by stealth, a road from Dijon to Belfast.
>
> (*Do or Die!* 1994 (4): 1)

A number of activists linked to EF!, some homeless, established a camp on the hill, in February 1994:

> I had been kicked out and I was crashing around people's houses so I thought I might as well go to Bath....[I] lived in a young homeless hostel in Bath and then four of us moved to the hill...set up a camp, didn't know what we were doing but set our tents up there. [On the] first day of construction ten or so people [were] ready to do stuff and went down and locked onto machines.
>
> ('Clare' in interview)

Former Twyford Dongas and other experienced road protesters poured into the camp, and a vigorous protest was mounted. On 14 March, when construction was supposed to start, 'workers [arrived] to find activists D-locked to machinery and sunbathing on site-office roof' (*Action Update*, 'Dijon to Belfast Euroroute', 1994). The next day saw three arrests, but the contractors managed to work for half a day. 21 March saw the occupation of the contractors' main office, with three houses taken back 'by protesters dressed as security guards' (*Action Update*, 'Dijon to Belfast Euroroute', 1994). *Do or Die!* (1994 (4): 1) noted:

> Site invasions are daily and protesters are presently occupying 8 acres of woodland and a water meadow [to save them] from imminent destruction. There have been office occupations, eviction sieges, mass arrests, 60-foot [high] 9-day tree-sits, mothers and children's direct action days, free festivals,

> burning bulldozers, marches (one with over 1,200 people) and every other conceivable tactic...

Links between middle-class campaigners, like the novelist Bel Mooney, and activists were again noted by the media. Mooney's teenage daughter, writing a piece in the *Daily Mail* entitled 'My friends, the gentle eco-warriors', reported:

> When I heard about the New Age travellers camped in tents...and saw shaggy people wandering about, I admit I was sightly prejudiced....Now I know what they are really like – these protesters some call 'Dongas' after an African tribe. The camps are organised, the people friendly – never teasing me about my conservative privileged background.
>
> (Quoted in *Action Update*, 'Dijon to Belfast Euroroute', 1994)

John Vidal noted in the *Guardian* (17 May 1994, section 2: 2):

> The Honourable Rupert Legge, son of the Earl of Dartmouth, novelist, land-owner and a man with Etonian tendencies, sits by his son's tent. It's pitched in what, until last week, had been a modest suburban garden overlooking Bath but now resembles Beirut or Woodstock. There's a 15-foot high pile of bull-dozed trees, rubble everywhere and dreadlocked men are playing drums and flute around a fire. The titled one is appalled, but only by the destruction.

Indeed, as with Twyford and the M11, massive national coverage ensued. 'Full-time resistance' continued into the summer of 1994:

> [D]uring the winter months there was no need for actions as rain flooded the construction site...workers were laid off for a month until the (very well vandalised) drainage mechanisms could clear up the water.
>
> (*Do or Die!* 1994 (4): 1)

Numerous office occupations occurred, including raids on the contractors, the solicitors responsible for evictions and Alfred Marks, the employment agency which helped recruit the security guards.

Violence was a major issue during the campaign. The 'spiky' versus 'fluffy' debate was launched here. 'Keep it fluffy' was an appeal for strict non-violence, and protesters referred to the 'fluffy fields', areas near the construction site covered in downy botanical fur. Other activists felt that self-defence was more than justified in the face of security-guard violence.

On 19 June there was action at College Green in Bristol when John McGregor opened the Avon Green Transport Week. Tania de ste Croix 'crowned' him with a wreath and his words were drowned out by protesters. The cross-currents of protesters' diverse affiliations were illustrated by the presence, to my left, of Ian Bone, the foul-mouthed founder of the notorious anarchist group Class War, and, to my right, of Tracy Ward, actress and Marchioness of Worcester (Ian had long been a friend, and I had made Tracy's acquaintance. I forget, now, whether I introduced them to one another).

Actions related to Solsbury Hill were sporadic throughout 1995:

> The one year anniversary gathering in mid March '95 went off with a bang…an anti-CJA event on the Hill ended up with lots of fencing pulled down, trashed machinery and security thugs in hospital (I guess people had had enough of being used as punch bags).
>
> (*Do or Die!* 1995 (5): 18)

As at Twyford, action petered out eventually, but the anti-roads campaign rolled on. Local network links strengthened and anti-roads action continued into the 1990s in Bath and Bristol.

Glasgow

Local councillors in Glasgow have long been some of the UK's strongest advocates of the urban motorway, despite the comparatively low level of car ownership in that city. During the 1990s two new routes, the M77 and M74, were promoted, but opposition to them from linked EF! (UK) and local working-class protesters was fierce. In 1993 Glasgow EF! activists occupied road-building equipment and helped create the Pollok Free State at Barrhead Woods, the first Scottish anti-road camp. EF! remained highly visible during the campaign, emphasising the movement's distinctive identity, due in part to the participation during 1994–5 of Burbridge, EF! (UK)'s co-founder.

In the 1970s the National Trust for Scotland had given permission for the M77 to cut through the green space bequeathed to the people of the city by NTS's founder, Sir John Stirling Maxwell. Opposition had begun in the same decade, and in 1988 campaigners connected to the 'Glasgow for People' group attempted to make use of a three-month public inquiry to prevent the destruction of a much-loved green space. During 1993 the preliminary work was disrupted by Glasgow EF!, which took occupation of cranes and trees. Hit with money hassles just as 'two cranes spontaneously combusted', motorway construction was briefly halted, allowing campaigners time to regroup.

1994 saw the formation of the umbrella-group 'Stop the Ayre Road Route Alliance', made up of environmental pressure groups, including FoE, local community groups and EF! Strathclyde Regional Council, in its dying days before abolition by John Major's government, had been the target of action and council offices were occupied by activists. This camp went from strength to strength.

Network ties were built with diverse groups, according to *Do or Die!*, while raves, gigs and other party events helped accelerate activism. Working-class activists were more visible there than at previous sites of road protest, and the socialist group Militant helped mobilise local support. 'Glasgow has a long history of community resistance, from Red Clydesiders onwards, and Pollok formed one of the centres of the anti-poll tax movement...the M77 will add to the long list of civil unrest seen in this area.' (*Do or Die!* 1995 (5): 8)

1995 was a year of dramatic conflict. The first day of February saw Wimpy Construction move in to fell trees. These trees had been 'spiked' by activists, but 'the cutters would fell the tree [below] the line of spikes', seemingly 'oblivious to the dangers spikes would pose' (*Do or Die!* 1995 (5): 8). On 14 February there was a 'Valentine's Day massacre' when 300 police and security men surrounded the camp. That the police made use of the local school's playground for their assault incensed pupils, 100 of whom walked out of classes and broke through police lines. Twenty-six security guards resigned on the day, Militant convincing many of them to quit.

The pupils' activity in the campaign increased. Organising a union, they demanded two hours off lessons per day to protest!

> So successful were the children's campaigns that the local media began to paint free staters and Militant as manipulating the kids into unrest....In an area with a high degree of truancy, the pupils needed no encouragement, and became a valuable asset to the campaign as they organised direct actions between themselves.
>
> (*Do or Die!* 1995 (5): 9)

Several days later the 'To Pollok with Love' convey of cars driven by anti-road activists from across the UK came to Glasgow. The cars were erected into a monument and ritually burnt 'on the rain-swept motorway bed'. This ritual destruction of 'carhenge', casting a spell against motorway madness, was condemned by Greenpeace as environmentally damaging.

Two new camps were established in 1995. At one, Patterdown, the Scottish Minister for Trade and Industry threatened campers with a pick-axe, and in consequence was forced to resign.

During 1995, after more expense and more media attention, evictions cleared all three camps, and the road ran through. The campaign had, however, stiffened local resolve on matters of social justice, ecological concern and Scottish working-class identity.

> The No M77 campaign has brought about the long spoken about alliance between Green and Red and made it a reality. Many environmentalists began to see the campaign beyond wholly moral terms and saw the class and social

> implications of this fight. Militant defending the trees would have been unthinkable a few months ago.... The struggle carries on...
>
> (*Do or Die!* 1995 (5): 8)

M65

Activists from the all-important Manchester EF! who had been at Twyford and Claremont Road realised in 1994 that they had a motorway project 'literally on their doorstep', and a direct action campaign against the M65 extension was established (Begg in interview). Geffen noted: 'People from Lancaster and Preston...were to take the experiences [from Twyford and Claremont Road] back with them to set up the M65 campaign; they were interesting because they were...very much stepping in the footsteps of the M11 campaign.'

The campaign began in the leafy Cuerden Valley, where a 'village in the sky' was established to slow evictions:

> There were over 40 tree houses ranging between 30 and 70 feet in the air protecting both sides of the approximately 100 metre wide valley. High tension cable walkways connected all the treetop homes and crossed over the River Ribblesworth, that surged (due to the raincloud vortex over Lancashire) through the middle of the valley.
>
> (*Do or Die!* 1995 (5): 11)

The drama of eviction and elaborate resistance was played out on 27 April, when the usual army of police, security guards and sheriff's men arrived. Three hundred activists were on site to celebrate the Celtic pagan spring festival of Beltane. As happened at Twyford on Yellow Wednesday, the prospect of partying and ritual enactment here had swollen the modest numbers: 'with music all around and the forces of doom and destruction gathering...the compound was stormed by naked protesters who left Group 4 in an embarrassed quandary' (*Do or Die!* 1995 (5): 11). It took climbers nearly a week to dislodge the camp; as tempers ran high, walkways between trees were cut down, activists snatched away and the trees swiftly chainsawed.

In contrast to Pollok, the M65 action was initiated by outsiders; but, like Pollok, support from local working-class communities was here apparent. Tactics from Solsbury, Jesmond Dene and the M11 were developed further. Naked protest was not the only sign of anarchic cultural activity: buildings, most notably a former police station on the motorway route, were squatted and dramatically redecorated. The new 'eco' police-station, according to garrison activists, was instrumental in increasing the cost of the motorway by some £12 million. The M65 campaign led to tactical innovation and the training of more activists in techniques of disruption.

Newbury

Newbury was billed as the new 'Twyford' because the proposed bypass would demolish verdant Berkshire countryside. With the waning of the roads vision, Dr Brian Mawhinney in the spring of 1995 ordered a year-long delay in the £77 million scheme. Yet, in what was more-or-less his final act as Transport Minister, he gave the scheme the all-clear. A campaign of strong opposition was countered by pro-bypass campaigners from Newbury town centre, including local Liberal Democrat MP David Rendell, who believed it would reduce traffic congestion.

The 'Third Battle of Newbury' campaign against the bypass linked EF! (UK), local campaigners and FoE. The 'first' and 'second' battles had been seventeenth-century skirmishes in the English Civil War, providing a useful historical icon for campaign reference. Newbury was the first direct action anti-road campaign to receive strong overt support from FoE nationally, a shift in policy supported by their (then new) Director Charles Secrett.

In contrast to other campaigns, the Newbury direct actionists made the first move. On 3 January 1996 protesters blocked the Reliance Security encampment. The contractors, Blackwell and Mott McDonald, refused to work without security guards. Similar actions slowed work during the first week of planned construction. By the end of the month 300 arrests had been made. In contrast to the tiny number of activists camped during the early days at Twyford, direct action mass protest was thus apparent from the first at Newbury. EF! (UK)'s *Action Update* observed:

> The last four years almost seem a rehearsal. People and skills acquired since Twyford are now coming through. As our resistance has captured the national imagination, the state is taking on the Environmental Direct Action Movement to try and break us.
>
> (*AU* 1996 (24): 1)

Around thirty anti-road camps were established on the nine-mile route, and by April arrests had risen to over 700. Such figures show how activism had grown: at Twyford mobilising a handful of individuals to maintain a single camp had been difficult. Tactics used both by activists and by their evictors also transformed with dizzying speed:

> They adopted tactics of chaos, dotting around the route, evicting whole camps, half camps, having night time cordons, sending the police on in advance, making ambushes, and varying their times of arrival. Every eviction was different.
>
> We developed various tactics and innovative defences where, as well as the basic walkway and treehouse networks, there were tripods, platforms,

> tunnels, scaffolding poles extending from the tops of trees and various lock-on techniques; and each of these worked with varying degrees of effectiveness.
>
> (*AU* 1996 (26): 1)

Although all the camps were removed by the authorities during the spring of 1996, and the security presence reduced accordingly, during the summer new direct action harried construction workers, with tunnels being dug, fences felled and work disrupted. FoE launched legal action to halt construction, claiming that work would destroy the habitat of a rare snail, Desmoulin's Whorl. EF! (UK) optimistically noted: 'It could be that the Tory slugs are defeated by 3mm snails and the bypass cancelled!' (*AU* 1996 (28): 1). It was not to be.

> As the last tree hit the ground, most of the security guards cheered. I didn't understand this; even without understanding our perspective, it meant they were now out of a job....As for us now, we sat down. We weren't enraged or furious. We were subdued. Wiped out. It's a weird set up where those who care most about environmental destruction are those who see so much of it first hand.
>
> (Merrick 1996: 115)

A rally to mark the first anniversary of work at Newbury was held on 11 January 1997, drawing around 800 people to the Middle Oak site. After listening to Charles Secrett and other speakers, part of the crowd broke down fences and occupied a work compound. Despite the use of mounted police to control protesters, by the evening 'a tip truck, portacabin and the crane were set alight' (*AU* 1997 (35): 1), much to Secrett's disquiet. As he disclosed in interview:

> The thing we part company over is violence. We won't tolerate violence...living in our country with our type of government. I don't think there is a justification for violence. I define violence in terms of people; when you come to property, it is a bit trickier. There are certain circumstances when certain types of damage to property, um, both technical and actual...[like] pouring sugar into a bulldozer [which is] going through an SSSI – I [don't] have a problem with that....[But I don't agree with] setting fire to a bulldozer, as happened at Newbury. You can't come into an event like that...hitting a BBC cameraman just because he was filming what was going on, throwing rocks....I might not pass comment if it was a small group of people taking contained action, or I would say that is not the kind of thing that FoE would support, but I can understand why others were doing it and why they felt they were morally justified in doing it.

Ultimately, however, and despite the largest anti-road campaign to date, Newbury is ringed with concrete.

Reclaim the Streets

The 'No M11' campaign refounded Reclaim the Streets (RTS) in February 1995 and initiated a series of increasingly ambitious interventions. The Nigerian Embassy was targeted as part of the protest against Shell's exploitation of Ogoniland and the repression of environmental activists in the country. The 1995 Motor Show was also an arena for protest. Above all, though, RTS worked to construct ever-more ambitious street parties.

Street party 1 was held in Camden Town in the summer of 1995. The streets between Camden Underground and the market are pedestrianised (and all but traffic-free) on a busy day. The area is a magnet for young people seeking quasi-hippie clothes, bootleg rock-concert music and various trinkets.

The geography, both physical and social, made it too easy. Where was the challenge in stopping some Sunday buses in north London's premier shopping zone for dreadlocked students, teenagers and alternative tourists? The significance was, of course, that RTS had launched an adaptation of earlier street actions within a territory where success was guaranteed. Confidence and experience gained in Camden allowed for more impressive future interventions.

The location of a street party, prior to the event, is known only to a handful of individuals. Publicity material draws participants to a meeting place whence they can be redirected to the actual venue. Before street party 1 we met at the Rainbow Centre, a large gothic church squatted by various anti-road activists and others. A tripod was erected to block the traffic; children's climbing frames were brought out and a montage of two crashed cars was left in the street. Several hundred participants listened to sound systems, chatted, ate, drank and made merry.

Street party 2 was held later in the year at the Angel, Islington. This, too, was a relatively soft north London target. While the road traffic was substantial along the A-route, the social geography again brought hippified youth and nostalgic 30-somethings to shop at the local market's stalls. The event was bigger than that at Camden, and culminated in dramatic violence when the police cleared the area in the early evening.

Street party 3 was remarkable in terms both of the nature of the territory and of the number of participants. In July 1996, 7,000 people occupied a stretch of motorway in west London in an action initiated by RTS and supported by other EF! (UK) groups (Anon. in interview; and participant observation). The event was planned with great care, and less than ten individuals knew the ultimate location (interview accounts). A tripod was erected, blocking the motorway, and thousands of participants were redirected from a meeting point at Liverpool Street station, east London, via the District (underground) Line. The police presence was too small to prevent the street party, which mobilised more individuals in illicit activity than has any other anti-road direct action protest event.

A biodegradable network: Earth First! 1992–4

Many of the founding activists moved to new campaigns relatively soon after the EF! (UK) network and the anti-roads movement of the 1990s emerged. The two EF! (UK) co-founders left the UK in search of 'wilderness', with Burbridge settling in Sweden and Torrance leaving for Tasmania. Zelter moved back into peace-movement action. Marshall, in turn, became more closely identified with campaigning video work for the group The Undercurrents (Marshall, Torrance and Zelter interviews). After the Brighton and 1993 Melton Mowberry Gatherings, more militant activists sympathetic to sabotage repertoires became disillusioned with the emphasis on NVDA (Gandalf 1993; Molland in interview). Garland left the movement after a lengthy dispute in Oxford EF! He and other militants promoted prisoner support and retained contact with EF! (US) radicals (Garland in interview). Eventually a number of former EF! (UK) militants became active in the campaign against live animal exports, initiated by local campaigners and animal-liberation activists, that raged at ports such as Brightlingsea, Dover, Plymouth and Shoreham ('Mix' in interview). Noble helped to organise a fruitarian and raw-food network.

New activists became involved (see Chapter 5). A few relatively large and stable local groups dominated the network, producing *AU* and organising national Gatherings. In 1992–3, Oxford EF! had fulfilled this role. From 1993 to 1995 the mantle passed to Manchester, a large student-based group which, like Oxford, was perceived as hostile to militant repertoires and sentiments (Garland and 'Mix' interviews).

All of these and later editorial groups had access to office facilities, a resource often lacked by smaller, more ephemeral, EF! (UK) groups. Oxford was able to use an environment centre in the east of the city, while Cardiff and Manchester had access to student union facilities. Such resources have provided positions of relative power, even in a deprofessionalised and biodegradable network. Throughout EF! (UK)'s existence, groups, both 'moderate' and more 'militant', with a strong resource base have influenced the direction of the network in terms of repertoires, organisation and issue-focus.

Between 1993 and the autumn of 1995 EF! (UK) increasingly dissolved or biodegraded into a wider anti-roads movement. Typically EF! (UK) local groups would initiate or support an anti-road camp to the exclusion of other issues. Green Man EF!, Bath EF! and Camelot EF! merged, respectively, into the No M11, Save Our Solsbury and Friends of Twyford Down anti-road campaigns:

> What was interesting was that EF! was a presence at Twyford Down but not at the M11 and subsequent road campaigns. You knew the people were EF!ers who you knew through the EF! network, but EF! never had a visible presence at the M11. There were no EF! banners or anything like that – it wasn't Leytonstone EF!
>
> (Geffen in interview)

Such biodegradation was accelerated by the politicising of youth dance culture threatened by the proposed 1994 Criminal Justice Bill (see pp. 125–6 and 137–9).

Identity and militancy: Earth First! 1995–8

During this period, repertoires of militant action became more acceptable within the network; EF! (UK) sought to reassert its identity, and relations with environmental pressure groups improved.

The Criminal Justice Act (November 1994) criminalised many aspects of non-violent mass direct action, decreasing the relative 'cost' of militant repertoires. A report on the May 1995 national EF! (UK) Gathering noted: 'Sabotage was also discussed. The opinions were diverse, but with the CJA now potentially threatening 3 month jail sentences for peacefully sitting on a digger the appeal of sabotage will obviously increase' (*AU* 1995 (16): 2).

In March 1995, Cardiff EF!, another student-based group, took over the production of *Action Update*. The number of issues produced doubled to twelve a year and it became more militant in tone. Typically, issue 15 reported the jailing of ALF activist Keith Mann, noting: 'His "crime" was criminal damage to 3 meat vehicles and escape from custody. Keith did not harm life, in accordance with the Animal Liberation Front's philosophy of nonviolence to all things.' *AU* 16 reported police raids on *Green Anarchist* in a similar tone. Militant sympathies increased with the creation of a London-based editorial group in April 1996. Self-defence was condoned, militant international campaigns including support for the US MOVE organisation and Bougainville Revolutionary Army (see pp. 7 and 183 respectively) were advocated, a concern with police and intelligence service repression articulated and examples of 'ecotage' publicised.

An autumn 1995 Gathering outlined plans to reassert EF! (UK)'s dissolving identity and to draw attention to the need for wider environmental direct action beyond the anti-roads movement (*AU* 1995 (21): 3). A national EF! (UK) action, the first seen since the rainforest actions of 1992, was planned to take place at Whatley Quarry, Somerset. Quarry expansion threatened the Mendip Hills and stone from Whatley was being used for road construction. Thus as well as attracting anti-road activists, the action was intended to emphasise the fact that direct action repertoires could be utilised against a variety of targets. The action, held in December 1995, attracted over 500 activists and much media attention.

Combining overt NVDA and covert ecotage, it reasserted EF! (UK) as a mobilising force.

Paradoxically, given such increasing militancy, relations with environmental pressure groups and the Green Party improved. In July 1994 Greenpeace and FoE had

> joined together in a unique action on the M65 construction site. The 'Rainbow Deconstruction Company' turned up...with two diggers... followed by a trailer full of tree saplings. The sixty strong work force then turned the scarred Cuerden Valley into an almost peaceful garden with creative road blocks, tree planting and stone mosaics. It took security guards at least an hour before they realised it was their turn to sit on the diggers!
>
> (*AU* 1994 (12): 6)

Later in July, Greenpeace activists chained themselves to concrete foundations at the second Severn motorway crossing, delaying work for twelve hours. Two national meetings were held between Greenpeace, FoE and the anti-roads movement, with activists from EF! (UK), RTS and Road Alert in attendance (Freeman interview; *AU* 1995 (21): 3). In September 1995, FoE Director Charles Secrett, accompanied by senior figures from his own organisation and from Greenpeace, camped at the 1995 Gathering and held discussions with EF! (UK) activists (Secrett interview). In a break with earlier FoE policy, he sought links with the EF! (UK) network. Discussion centred on why relations had been poor in the past and how informal links could be made between the organisations.

> It was about making face-to-face contact,...We are not ignorant or arrogant [enough] to think that we can do everything on our own....We had a *kerfufle* over the Newbury rally, but that was not slagging off EF! or direct action, [but] just hot-headed individuals. We were very precise and careful about what we said. This sparked off a debate in EF! about when violence is justified. Not everybody agreed with us, but they could see where we were coming from.
>
> (Secrett in interview)

Resource implications, given the high profile of the anti-roads movement, may have motivated an increasing pressure-group sympathy for NVDA. According to *AU*, some (EF! activists) felt that FoE was trying to use them as a vehicle to recruit in the 'youth market'. Others argued that alliances might be made at the local level and suggested that FoE activists might support the forthcoming Whatley action. FoE became strongly involved in the Newbury Bypass campaign, after earlier rejecting participation at Twyford and the M11. Equally, Freeman (in interview) argued that resources, particularly in terms of access to research information, were donated by FoE to EF! (UK) and the anti-roads movement. In 1993 the Green Party, in contrast to its earlier opposition, had created a Committee of One Hundred to promote NVDA, which worked closely with EF! (UK) in organising and resourcing the Whatley action (Begg in interview).

Militants, though, continued to challenge these organisations. Typically, *AU* 27 contained 'An open letter to Greenpeace UK' from South Downs EF!, condemning the organisation's criticisms of German anti-nuclear militants. In turn, as described above, Secrett condemned anti-road militants for rioting at a Newbury rally in January 1997 and destroying construction equipment (Secrett in interview; Moyes 1997: 3). While FoE had come to sympathise with mass NVDA, disillusioned 'militants' such as Mix and Molland had re-entered the EF! (UK) network. By 1996 mass NVDA was increasingly combined with repertoires of mass sabotage. During street party 3 a giant carnival figure surrounded by thousands of activists provided cover for unknown individuals using pneumatic drills to dig holes in the road surface (Bellos and Vidal 1996: 7).

'Swampy fever' and media spectacle

Numerous other anti-road campaigns were waged, the majority using conventional pressure-group tactics to produce 'quiet victories', aided by the 'noisy defeats' outlined above (Stewart in interview). The earliest of these successes came at Oxleas Wood in 1992. People Against the River Crossing and the local FoE branch had opposed the scheme for years. Secrett remembers the national FoE working on the campaign in the early 1980s (Secrett in interview). The east London River Crossing would have cut through a wood claimed to be 8,000 years old and demanded the demolition of hundreds of homes. An alliance was constructed between FoE, the World Wide Fund for Nature, EF!(UK) and others. This mixture of direct action radicals and conventional pressure groups collected a 10,000-name 'Beat the Bulldozer' pledge by individuals committed to non-violently resisting the road. The government withdrew the scheme (Torrance in interview; Stewart *et al.* 1995). Campaigns against the widening of the M25 to fourteen lanes, the M1 and M62 link in Yorkshire and the Salisbury Bypass were all won using the threat of direct action and conventional pressure-group lobbying.

The noisiest defeat of all came in January 1997, with evictions at the A30 site marking a dramatic end to a series of permanent protest camps that had been established for several years in east Devon. The campaign saw the press deification of young tunneller Daniel Hooper, or 'Swampy'. The media celebrated roads protest and crystallised the actions of thousands of grassroot campaigners into the form of a single personality. Swampy became an icon, better known, at least briefly, than many TV presenters and cabinet ministers.

Such celebration was a double-edged sword. Daniel Hooper had not asked to be a media star, and, while he attempted to exploit his fame for the benefit of the protest movement, he was clearly embarrassed by his role. Both EF! (UK) and the wider anti-roads movement had been largely successful in combating personality politics prior to the 1997 evictions. The rejection of personality was seen as a means of making protest accessible. Media fame brought the theme of protest to millions of individuals, but in doing so raised its exponents to the status of the unique. In short, direct action was transformed from the 'real' to a media spectacle, something for heroes rather than adults and children in local communities.

Although road protests continued after January 1997, these Fairmile, east Devon, evictions marked the end of a long phase of protest. The programme of road construction had been scaled down and new targets attracted green direct action attention. In 1997 there was vigorous action against construction at Manchester Airport. Localised protests have continued, and at the time of writing (1998) the Birmingham Northern Relief Road has attracted a new wave of camps. Meanwhile, applied genetics, strip-mining and building on the greenbelt have become campaign targets.

> From bypasses to semis: direct action campaigners say the time has come to shift from trying to stop roads, where they have had propaganda and policy success, to blocking new-house construction programmes on greenfield sites....Paul de Luce, an Earth First! supporter and veteran of the Twyford Down and Newbury Bypass protests, said: 'Direct action against house building in rural areas is a natural progression. I'm expecting it to be more frequent and larger than with roads because it will have more support from local people.'
>
> (Schoon 1998: 1)

On the housing issue, FoE's Director has argued that, 'with the CPRE, we have a wonderful alliance, [from] one of the most conservative groups through to FoE through to the direct lot at the other end of the tactical spectrum' (Secrett in interview).

Although previous roads programmes have been hampered by recession, with construction limited by the economic crisis of the early 1990s, protest has made its contribution to the reduction in road building:

> That terrible scar in the Hampshire countryside through a gentle hill called Twyford Down which used to stand among the surrounding woods is a memorial not only to that terrible loss but to a series of subsequent victories....'We don't want another Twyford Down' has become the slogan of both the anti and pro roads movements. Talk to the British Roads Federation, and their spokesmen (they are invariably men) are conciliatory, apologetic almost. No longer do they have visions of concreting over Britain. They

> realise their best hope lies in slipping in a road here, a by-pass there, with the minimum of fuss.
>
> (Wolmer 1997: 41–2)

As social movement theorist Sydney Tarrow (1994) observes, the legacy of mobilisation is not just policy change but the development of new tactics, the recruitment of new activists and changes in popular consciousness. Activist involvement and the transformation of personal consciousness are the focuses of Chapter 5.

5 Activist identity

> University just didn't seem important. I was already thinking, 'I want to be in the real world', listening to Jai spouting on about car culture, and thinking the world is going to die in so many years, you know. [So] what the fuck is a degree...? I went back to my rooms and packed my rucksack. I had been at university for a month, and I ran off and joined the Dongas tribe.
>
> (Lush in interview)

Introduction

Activists and academics alike are interested to know why, while most of the population remains apparently apathetic and uninvolved, some individuals become immersed in direct action. Many a time around the camp fire road protesters have speculated about why they are involved. Life-histories are recounted and the inhabitants of the outside world may be gently, or not so gently, criticised. Academics remain unsure of how activists take on their identity. Believing in a 'cause' is not enough. Millions of people in Britain are critical of aspects of government policy but, apparently, only a few thousand take strong forms of action. A sizeable minority of the British public is sympathetic to the goals of the anti-roads movement, supporting some limits to motorway building and car use, but relatively few actually trespass on urban routes in Reclaim the Streets actions, and fewer still take to the trees or tunnels. Many studies indicate a huge gap between the number of individuals who claim to support the broad aims of a movement and the tiny proportion of these who participate in campaign activity (McAdam and Paulsen 1993: 643). For example, Klandermans and Oegema (1987) noted that while 74 per cent of a sample of individuals supported the goals of a peace-movement demonstration, only 4 per cent of that total actually attended. While explaining participation in general has proven difficult, explaining why some specific individuals become activists, undertake intensive action and initiate mobilisations presents an almost intractable problem. My interviews with twenty-nine EF! (UK) activists are used here in an attempt to unravel some of the knots.

Activist involvement

Diverse theories have been used to explain why, while the vast majority of the population stays at home, a minority takes on an activist identity. Accounts of self-interest link class to participation, stressing the material benefits of involvement (see Steinmetz 1994). Historically, Labour movements have sought better pay and conditions; middle-class campaigns such as rate-payers groups try to reduce the burden on individuals of local taxes; while nationalist movements have fought to improve the political and economic status of their regions. Supposedly 'new social movement' activism has been seen as the product of the emergence of a new (predominantly middle-class) stratum which attempts, via environmental protest, to gain improved social status and greater economic power (Kitschelt 1985: 278; Mattausch 1989: 50; Steinmetz 1994: 183). Yet 'new-class' explanations have been harshly criticised. In addition to doubts about 'newness', the direct economic benefits of opposing road construction are not always obvious. Martell (1994: 130) concludes that while environmental movements 'may be class-based' they do not appear to be 'class-driven', suggesting that political, and particularly economic, gains for a specific 'new middle class' do not seem to follow from green demands. Steinmetz (1994: 183) suggests that better-educated social factions find it easier to become active than do other elements of a population. Equally 'new' middle-class groups may have greater access to the resources and spare time necessary for protest. Such a new class, in Bagguley's view (1992: 27), 'affords a "leadership role" to a wide range of social movements: new, radical, reactionary or otherwise'. Thus class does not just produce varied grievances: it enables activism in less obvious ways.

As the survey in Chapter 2 suggests, the class composition of green movements may be more varied than some researchers have suggested. Certainly 'the early green politics' was linked in the late nineteenth and early twentieth centuries to the Labour movement (Gould 1988). Kimber and Richardson (1974: 2), examining the early 1970s, noted the existence of 'activity in working-class areas such as the Swansea housewives' blockade of the United Carbon Black factory in February 1971, the unofficial and official activity leading to the closure of the Avonmouth lead and zinc smelter in January 1972, and the activities of groups like Clean Air for Teeside'. The experience of groups like MOVE in the US also indicates the multi-class and multi-culture nature of modern green activism.

Other explanations, using more personal traits, have been put forward (McAdam and Fernandez 1990: 1; Zukier 1982: 1090). Right-wing activism has been explained in terms of the existence of an authoritarian personality (Adorno *et al.* 1950; Hoffer 1951), while upbringing and other aspects of family background, and psychological factors have been examined with much use of elaborate statistical packages to account for the later tendency towards political participation.

Yet accounts that explain participation in terms of identity, and imply that identity grows from a single factor such as class or upbringing or whatever, are inadequate. They fail to indicate why the majority of surveyed individuals who, due to such traits, can be thus classified as potential activists fail to become involved (Mueller 1980: 69). Not all rebels at school become anti-road activists; so how do we explain the participation of the tiny number of individuals who establish anti-road camps and take part in EF! (UK) protests?

An alternative approach sees activist recruitment or the construction of an activist identity as processual or trajective, involving a number of factors. Direct action campaigners

> are expected to (a) have a history of activism, (b) be deeply committed to the ideology and goals of the movement, (c) be integrated into activist networks, and (d) be relatively free of personal constraints that would make participation especially risky.
>
> (McAdam 1986: 71)

Friendship networks may ease individuals into activism:

> Imagine…the case of a college student who is urged by his friends to attend a large 'anti-nuke' rally on campus. In deciding whether to attend, the potential recruit is likely to weigh the risk of disappointing or losing the respect of his friends against the personal risks of participation. Given the relatively low cost and risk associated with the rally, this hypothetical recruit is likely to attend, even if he is fairly apathetic about the issues in question.
>
> (McAdam 1986: 68–9)

Such involvement allows individuals to get involved gradually, intensifies peer pressure and provides new friendship links. Few individuals jump onto bulldozers or tunnel under construction sites when first involved; repertoires have to be learnt. Activism, even in its most serious form, is a method of performance that must be developed and improvised.

Equally, despite the mud, the cold and the threats of confrontation and arrest, direct action may be exciting or even enjoyable, encouraging the pursuit of more intensive activism (Nelkin and Pollak 1982: 78; Roseneil 1995: 57–9). Such experience may lead to personal change that reinforces a new collective identity, whether in the form of Melucci's 'we' or via 'green agency' as discussed in Chapter 1 (Della Porta 1995: 162; Fantasia 1988: 17; Mueller 1992: 5).

Despite appropriate network links, personal political sympathy or even many years of experience, not all movement supporters engage in intensive or 'high-cost' activities, such as living in protest camps, sabotaging machinery or risking life and limb sitting in trees under attack from motorway builders. Personal availability is the key factor: those of us who are short of time are unlikely to be significantly 'active'. Mundane commitments that absorb time, energy and finance are a barrier to participation. Potential 'activists' are comprised mainly of a small minority of the politically concerned who lack work and family commitments (McAdam 1986: 85). Activists are predicted to be young and free of commitments such as caring for children, elderly parents or other dependants. They are likely to be unemployed, or students or, more rarely, self-employed with access to discretionary time (Wiltfang and McAdam 1991: 998). While unemployment may give individuals a reason for protest, it also resources them with the time in which to protest. Equally, having free time and a grudge are in themselves insufficient to constitute 'the protester': also required are a cycle of gradual involvement and strengthening network ties.

In line with the critical realist approach adopted here, an open question put to individual interviewees asked how they became involved as activists. The characteristics just outlined emerged from the examination of interview transcripts. In short, the research did not presuppose defined trajectories of involvement and then attempt to classify their exemplars; rather, a process of evolving activist identity along such trajectories is what quickly became apparent.

Activist trajectories into EF! (UK)

Accident, friendship, personal conviction, pleasure and political calculation were just some of the factors cited as influential by the twenty-nine interviewees from EF! (UK). Despite such diversity, common features were apparent. All activists, from the two cohorts into which the twenty-nine were grouped, had moved through a process of attitude change and increasing activism. Most of the first cohort of fourteen interviewees were founder-members of EF! (UK) in 1991–2 who had been greens for some time. Sympathy with green ideas was followed by membership of a green movement group, such as the Green Party or FoE, or a peace or animal-liberation group. Membership may have taken a passive form, with the individual doing no more, perhaps, than receiving newsletters or sending donations. Generally, for the interviewees, such passive membership was followed by more intensive activism. Most had been activists in other groups before undertaking such a role in EF! (UK). While apparently unmotivated by career considerations, they could nevertheless be seen as political entrepreneurs who consciously mobilised resources and made effective use of repertoires of action.

Contrasts were apparent with the second cohort, who became key activists in EF! (UK) after its initial mobilisation These 'follow-on' political entrepreneurs usually had been active outside of the green movement, constructing a green's identity as part of a process of EF! (UK) participation rather than possessing it as a prerequisite. Thus early activists were greens who came to EF! (UK); later activists were more likely to have entered the green movement family as a result of anti-roads involvement.

Environmental and political framing

Most of the first EF! (UK) activists had been sympathetic to green ideas before joining a green group. In fact, 10 of the first set of 14 interviewees felt that they had become environmentally aware before participating in the green movement. 'Environmental awareness' included an appreciation of the rural environment, concern for other species and an awareness of environmental problems such as pollution. Several interviewees cited childhood experience as influential in this respect, while two noted memories of environmental change/disaster witnessed as children:

> There was a fairly untouched valley over the back of the council estate, called Ore Valley, where they have really fucking destroyed it and built another fucking council estate there now...and I used to play there a lot and there were streams and whatever running through there.
>
> (Torrance in interview)

Garland remembered reports of the *Torrey Canyon* disaster when an oil-tanker accident led to much loss of wildlife and television images of oiled sea-birds. Such incidents of suddenly imposed grievance as the nuclear accidents at Chernobyl and Three Mile Island, or the wreck of oil-tankers such as the *Torrey Canyon*, distantly and perhaps not whole accurately remembered, may dramatise environmental concern so as to 'assist political identification of the nature of an issue, the situations out of which it arises, the causes and effects, the identity of the activities and the groups in the community which are involved with the issue' (Hannigan 1995: 46. See also Staggenborg 1993).

Marshall felt that erosion of a much-loved and mythologised landscape, experienced as a child, had influenced adult commitment:

> It was a way of life that has almost, now, completely died out and...I was aware it was dying...it was very clear in the Forest of Dean. Dennis Potter has written very good stuff about the way the outside world – really, after the war, especially in 1950s and the 1960s – started just crashing in. First of all radio and then television came in, and then cars, car ownership and how that broke things up; and then the Forestry Commission came in and completely fucked the place...cut down huge areas of the original oakland and oak forest which, of course, had the sheep in as well....That all went to be replaced by conifer plantations. All of this was very upsetting to me at the time.
>
> (Marshall in interview)

Such an account fits with Welsh's view that 'transgression of a place cherished in a childhood memory' may fuel later activism (1994: 11).

The importance of more routine childhood contact with the environment was mentioned by other interviewees:

> I did quite a bit of my growing up in Yorkshire, in the Dales....I wouldn't call it wilderness but...in this country it's quite a nice wild area and animals are my big passion in life. I wanted to be a vet when I grew up, and all that kind of stuff.
>
> (Tilly in interview)

Another activist saw her childhood as having been a 'green' one based around a strict fundamentalist Christian lifestyle that forbade visits to the cinema and demonised much popular culture, and was lived-out in as self-sufficient a fashion as possible (Noble in interview). 'Mary', who was politically 'aware' prior to discovering any personal commitment to environmental issues, was more sceptical about the formative power of childhood green lifestyles, sectarian or secular:

> It could be really funny, but it is the only way I can think of answering it. Being taken on long soaking walks in Derbyshire in the pissing rain and mud, and being told: 'Cor, can you see that bird up there?' 'Where? Why can't I see it? Where is the car?'...Environmental awareness? Not really as a child, no.
>
> ('Mary' in interview)

School was seen by several individuals from the first cohort as influencing later 'green' identity. Durham cited the importance of A-level studies in introducing him to the German Green Party. He had also become interested in vegetarianism after he had 'dressed as Ronald McDonald' in a sixth-form college presentation on vegetarianism: 'I had to run on at the start and sing the Beef Patty Song.' Begg, too, saw school influences on environmental socialisation as significant. He also cited children's television and books as important:

> Things that strike me as being quite influential, now, are some of the children's programmes that were on, like the *Magic Roundabout* and *The Clangers* – an awful lot of ecological messages in *The Clangers*. Um, children's books, *Dinosaurs and All that Rubbish*....Whenever I come back to that book I feel a little pang of my childhood so there. So it certainly made an impact on me at the time.
>
> (Begg in interview)

One activist noted the value of urban environments:

> Now environmentalism is based on looking at or rather [being] obsessed with ecology and green things, and I share that very much. But, you have to realise, for most people the urban environment *is* their environment and the constant demands of capitalism for change, for change and for writing off – you know, you can write off tax by this refit, refit, refit, smash down, buy it up, get a new chain, build a chain! The way that everything turns into the same homogenised environment is, I think, an aspect of environmentalism that...started bringing me in.
>
> (Marshall in interview)

In contrast, 8 of the 15 interviewees in the second cohort emphasised their perception that personal environmental concern had grown after an earlier process of politicisation:

> I stayed with the campaign and went through a green politicisation that hadn't really been there before...it is still going on....I hadn't really marked out roads as any kind of political issue at that point. I knew in a general sense that it was wrong to go around covering the environment in concrete, but there was also this sense of 'What about the fact that schools are being closed down and hospitals are being closed down, and there are so many homeless people?'...This seemed to be a priority over this fairly abstract thing about roads....Because roads have become a political issue, it is obvious now, but it wasn't marked out as a contentious issue. Certainly for me, in 1993, it took a good couple of months to get my head round it.
>
> (Anon. 'Green' in interview)

The remaining seven second-cohorters echoed the accounts of the first cohort, noting variously their perception of the significance of a rural family background, experience of environmental problems and the influence of education. 'Lorenzo' remembered how he had completed a school presentation 'on endangered species' at the age of 6. Freeman at the age of 5 had been upset by a road being driven through deer-filled woods near her home. 'Clare' remembered her childhood as one of 'playing in the countryside, chasing the rainbows and setting up tree-houses'. Hunt observed:

> I suppose I just had an attachment to rural and countryside issues because of my upbringing. Partly because we lived on the margins of the countryside, cows escaping on to the bungalows where we lived, and explored the countryside around Chippenham and north Wiltshire. And my grandfather was an amateur naturalist, so that perhaps made me more receptive....I was

> taken along to fox hunts and things like that, so I could see it from both angles.

All twenty-nine interviewees, by varied routes, came to see environmental issues within a political frame, articulating the key green themes of non-violence, decentralisation and social justice together with environmental concern, and conceptualising them within a framework of political explanation. For most of the first cohort, environmental concern was followed by political framing, a sequence that was reversed for almost half of the second cohort.

Of the first cohort, Marshall, 'Mix', Noble and Tilly all linked politicisation to their 'rebellion' against family or school authority. Noble noted:

> I think, probably, it goes back to my father, because I grew up in Gloucestershire with a slightly mad, eccentric father....[My parents were] Christian fundamentalists....I wasn't allowed to indulge in things like cinema and comics, which of course I did on my own....So I learnt to be a rebel pretty early on.

Marshall claimed that his mixed-class background and socially disabling experience of public school had made him 'a polite rebel':

> I already hated the idea of being inside a hierarchy, taking orders from my boss....I have held very, very few jobs I haven't been sacked from, and that's something you might well find is a common string for people who get involved in direct action.

Libertarian features of childhood were also cited as a possible influence on later political activity. Garland stressed the value of living in St Ives and Glastonbury, with their strongly sub-cultural 'arts' and 'hippie' influence, while claiming: 'The value-system of my mother pushed me....I grew up totally in that kind of radical background.' Begg noted the influence of his liberal schooling:

> I went to a very bizarre school called Trinity that was a Catholic state-aided school, with no school uniform, first-name basis with teachers. Teaching focused on self-discipline, learning the subjects you want to learn....I think that had a lot of influence on me. I know a lot of other people at Trinity [who became social movement activists]....Indra Cooper went on to be a member of the editorial committee of *Peace News*.

Several interviewees perceived an activist or political culture as part of early family life. 'Mary' felt that her participation in anti-Gulf War actions 'really (brought back) a lot of my memories [of] when I was a kid. I spent a lot of my childhood pushing or carrying CND banners around London on big demos and going on trains to demos and stuff, you know, that brought a lot of that back.' The parents of three of the first cohort had been local-government councillors, of whom one had been a Liberal Democrat mayor and another the Labour vice-leader of a northern metropolitan council.

While childhood experience may have influenced their political awareness, activists also stressed the importance of certain green texts. Tilly noted: 'The first influential book for me was... *Blueprint for Survival*. In the 1970s, that was the first thing I read, and that really turned me on.' Noble and Zelter also referred to *Blueprint*, Collins to *Seeing Green*, Durham and Molland to *A Green Manifesto for the 1990s*, and Begg to *Nuclear Power for Beginners*, while *Deep Ecology* was of vital important to Torrance. Such texts 'framed' environmental concerns in political terms:

> I read the book [*Nuclear Power for Beginners*] and I can remember being quite struck with it, quite taken with many of its arguments, but very startled when it started to say capitalism was necessarily destructive of the environment, and I was thinking 'Surely not', and that planted some seeds in my mind.
>
> (Begg in interview)

Friendship links, as expected, were often important in initiating and enhancing such political identity formation. Torrance noted:

> Jake and I were really good friends....I got to know Jake when I was at college....We picked up a deep-ecology book from the Hastings' town and country fair. It went from there, really, and really, for us, we began our green political education then. We really bounced off of each other and really clicked, and we had these massive six- or seven-hour political discussions.

The second cohort of activists noted similar influences. Anon. perceived his anarchist politicisation to have stemmed from 'getting in trouble at school and that nauseating old route'. 'Mark' noted how friendship ties had made him politically aware and propelled him into the Labour Party. Brett noted, like Garland, the influence of 'hippie parents', childhood curiosity about the counter-culture and the cathartic effect of being assaulted by the police while a bystander at a Poll Tax demonstration. Several activists from the second cohort noted the activist/political culture of their parents. 'Lorenzo''s parents were 'bureaucrats in the EC Environment Commission', while both Allen and Lush's fathers were workplace shop-stewards. Lush also noted: 'I remember my mum really going off on one about nuclear weapons and that had a big effect on me.'

Political beliefs shifted and changed radically for both cohorts. Equally, EF! (UK) and anti-roads activism were justified by reference to different traditions. For Laughton and Marshall, political socialisation initially meant the adoption of a libertarian right-wing agenda. For Allen, Green and 'Mark' (from the second cohort) socialism was later fused with green politics. For 'Mix' (first cohort), 'Jazz' and the two Anons (all second cohort), anarchism was the political frame of reference, and continued to be so after involvement in EF! (UK) and the wider anti-roads movement. Molland had been a Liberal; others from the first cohort joined the Ecology/Green Party.

Network membership

Interviewees were asked about participation in green groups and other political organisations prior to activity in EF! (UK). All of the first cohort had previously been members of green groups. Noble and Zelter had become active in the early 1970s at the time of the publication of *A Blueprint for Survival*, when environmental concern was high in the UK (see Chapter 2 *passim*). Tilly had been inspired by the success of the German Green Party in 1983, as well as by reading *Blueprint*. Younger activists cited the wave of environmental concern in 1988–9 as crucial (see Chapter 2). Even those who had been active outside of the UK had been involved in green groups. Marshall, for example, had joined the Australian rainforest movement. Eleven had been members of environmental pressure groups. Of these, 5 had been FoE members and 4 Greenpeace members. The Henry Doubleday Research Association, Plantlife, the Soil Association (groups concerned with organic agriculture) and Survival International (a native people's civil rights organisation) also came into this category, each having had a single member of the first cohort. The majority (9 of the 14) had been Ecology/Green Party members, while 1 of the 9 had also been involved in *Green Anarchist*. Two other interviewees from the first cohort became involved with *GA* after joining EF! (UK). Molland was imprisoned because of his involvement as editor of *GA*'s diary of direct action events.

Most (11 of the 14) had also been members of the peace movement. 'Membership' ranged from living for many months at Greenham Women's Peace Camp to simply joining CND. Three interviewees had participated at Greenham, while one male interviewee had lived at Molesworth's (mixed-sex) Peace Camp. Seven had been members of anti-Gulf War groups.

Four had been members of animal-liberation groups, including the ALF Support Group and the Hunt Saboteurs Association, which practise disruptive direct action, as well as the League Against Cruel Sports, which does not.

Two had been members of anti-nuclear power groups, including Stop Sizewell B and WISE.

Thus a majority of founding EF! (UK) activists had previously been members of green political organisations, environmental pressure groups and social movements. The majority had been involved in the peace movement. Several had also been active in animal liberation groups. Surprisingly, only one had been a member of an anti-roads group. This may reflect the decline of anti-roads campaigning after Tyme's actions at public inquiries in the 1970s. Outside of the green movement, memberships of two Marxist groups, a local anarchist group, an anti-poll tax union, the Liberal Democrat Students and the Federation of Conservative Students were noted. Thus, five individuals had been members of non-green political movements, albeit of diverse affiliation.

Of the second cohort, a large minority had not joined a green group prior to membership of EF! (UK), and many saw their anti-roads activity as a gateway into the green movement. The remainder had been active in environmental pressure groups, green political organisations like the Green Party or *GA* or 'green' social movements. Again just one, Geffen, had prior involvement in anti-roads action. Nearly one-third had been members of non-green political organisations, including the Socialist Workers' Party, 'anarchist groups' and the Irish Marxist organisation People's Democracy.

Prior activity

All of the first cohort had been 'active'. Most had acted as, of course, unpaid political entrepreneurs, mobilising resources via networks to build particular events or campaigns. Most had also participated in non-violent direct action before becoming involved in EF! (UK) and the anti-roads movement. All fourteen had been highly active. The least involved, Durham, had participated in demonstrations against the Gulf War and, as a FoE energy campaigner, distributing free low-energy light-bulbs. Even the youngest activists, with the exception of Durham, had several years of previous activity and had acted as movement organisers.

In most cases such mobilising activity was multiple, long-term and involved the creation of new campaigning bodies or the adaptation of repertoires of action. Collins had squatted an office for the London regional Green Party and other social movements/cultural organisations. Garland had organised a Hunt-Saboteurs speaking tour across Europe. Begg had hosted a Green Student Network (GSN) conference at Leeds University, and also created a 'green' office for local groups in the city. Zelter had helped to found an organic-agriculture research group and had developed repertoires of innovative direct action. Noble co-founded and acted as spokeswoman for a group advocating a raw-food and fruitarian diet, enjoying semi-celebrity status including an appearance – while an EF! (UK) activist – on BBC's *Ruby Wax Show*. 'Mix' had helped to edit *Green Anarchist*.

Many of the first cohort had taken part in previous direct-action protests. All three women over the age of 30 had been at Greenham Common Women's Peace

Camp during the 1980s. Tilly had camped at the protest for a long period, while Noble and Zelter's activity was more occasional. Zelter initiated the Snowball Campaign of direct action for peace. Garland had participated in animal-liberation, anti-nuclear power and peace-movement actions. Marshall, as noted, had been active in the Australian rainforest movement.

Earlier involvement with direct action was associated with powerful and positive emotion. 'Mary' had led an illegal march on the House of Commons at the start of the Gulf War. For her, both the event and the subsequent news coverage of the action produced feelings of euphoria:

> We were done for obstruction of the highway but...when we were let out and we saw all the newspaper headlines, and it was front-page news, you know, these people holding this banner and marching through London in the middle of the night, um, and the number of arrests...that really kicked in and made an impression.
>
> ('Mary' in interview)

The ability to take on varied identities and learn new skills was seen as a product of intensive activism:

> There is no area that I can imagine which allows quite such a scope for doing different things....You have to be a journalist, you have to be an artist, you have to be a public speaker...you have to be very straight and very bent. You have to be able to do research and similarly you have to be a demagogue.
>
> (Marshall in interview)

Skills learnt in one campaign could be transferred to others. Thus activists tended to move from one sector of the green movement to another as the focus of their participation shifted with changes in wider green movement fortunes. Many had participated in the resurgent peace movement of the early and mid-1980s before focusing again on global environmental issues at the decade's end. Several were strongly involved in anti-Gulf War action. Clearly, pragmatic shifts of issue-focus within the green movement are apparent, with individual 'spill-overs' into new areas of campaigning as different issues are seen to have become more important. Activity outside the broad green movement family, such as participation in the anti-Nazi mobilisation of the late 1970s or campaigns against Poll Tax in the late 1980s, was not emphasised by the first cohort. The women's movement was an exception, and was seen as intrinsic to participation at Greenham.

Many of the second cohort had taken on similar roles, participating in direct action and working to mobilise resources. Some had participated in the green movement. Allen had mobilised community anti-toxic groups and also worked for Greenpeace. Freeman had acted first as a volunteer and then as a professional organiser for FoE, while others had been active in the Green Party itself. In contrast to the first cohort, several of the second had acted as political entrepreneurs outside of the green movement. Anon. noted: 'I was quite heavily involved in the Poll-Tax movement...setting up groups and getting people down to demonstrations, [and] involved in a lot of anarchist groups and Gulf War groups.' 'Green' stressed her 'involvement' with 'left-activist culture in the 1980s...things like Clause 28, the Alton Bill, anti-apartheid, CND, the March for Jobs, the GLC...student-loans campaigns, the Poll Tax. I think my politicisation occurred in the mid-1980s within that culture.'

Frustration and empowerment

Seven of the first cohort and six of the second cited 'frustration' with the existing green movement as a fundamental influence on their participation in EF! (UK) and the anti-roads movement. Existing green organisations were seen as relatively bureaucratic and ineffective. The opening words of EF! (UK) co-founder Torrance's interview illustrate this perception particularly well:

> I'd been involved in a wide variety of green groups, uh, from setting up with friends a local Greenpeace group and then getting involved in a Friends of the Earth group, and canvassing for the Green Party, and I had really gone through each of them, you know, in succession, really, getting more unfulfilled as I was going along and really feeling very deeply that there was something missing. I felt that I had a lot to give and I really fucking wanted to do something and [wondering]...why there wasn't anything inclusive set up, for a green grassroots in this country.

EF! (UK) was created by green movement political entrepreneurs, with the intention that it would be personally empowering and aimed in a strategically more radical direction than that of existing green groups. Such a motive does not, of course, account for the participation of other non-green activists in the anti-road protests of the 1990s.

> You get involved with Earth First! because Earth First! does something and these people don't do anything, and a lot of the things that drew people to Earth First! were things like non-hierarchy, no central office, no one telling you what to do. The idea, the idea that it is personal empowerment, like you can do whatever you want to do, [that] you operate within a consensus system. You don't necessarily do exactly what you personally want to do, but nor do you have a boss.
>
> (Marshall in interview)

The emphasis on direct action was seen as immensely satisfying by the founding activists:

> I suppose I had got disillusioned with the Green Party, as most people had, [I] felt it was too organised and too bureaucratic, and all the rest, and wanted a bit more direct action. So I started turning up at some of those demos and doing things, you know, directing a lot of energy towards showing people we just weren't going to put up...with, you know, the nonsense any more.
>
> (Noble in interview)

Collins observed:

> It was definitely tangible at the time....You may not stop the ship docking or stop the road project, but we all had a good time and it was exciting....I had never scurried around hedges at the dead of night trying to avoid searchlights and security guards, and it was incredibly empowering.

Collins argued that EF! (UK) action was having a powerful symbolic effect:

> [I]t was getting in the media, and it was getting attention, and this seemed like a publicity-rich way of doing things: it was a new label and in the media's view it was exciting. So, if you said things with an Earth First! hat on, you were far more likely to get quoted and noticed and make the news rather than a local Green Party thing.
>
> (Collins in interview)

There were clear exceptions from within the first cohort to this motive of overcoming frustration through empowerment. 'Mix', for instance, while satisfied with *GA*, felt that EF! (UK) and the anti-roads movement provided a vehicle with which to spread the demand for participatory politics and the revolutionary direct action he espoused. Durham, rather than being frustrated by his participation in FoE, was instead inspired by the enthusiasm of friends to join Manchester EF! Zelter, as noted earlier, saw her participation as 'accidental': it had been activist networking that brought her into contact with Burbridge and Torrance. Molland, for his part, was inspired by an ideology of deep ecology which, he felt, complemented his concern with animal liberation. All interviewees from the first cohort saw EF! (UK) as a means of promoting a politics of participation and applying direct action repertoires.

Many of the second cohort echoed the frustration of the founding activists. 'Claire' had helped create a local Greenpeace group, but was disciplined for trying to campaign on local transport issues. Hunt had also moved from a Greenpeace group to EF! (UK). 'Lorenzo''s thirst for direct action came after working for the World Wide Fund for Nature, attempting to protect sea-turtles, as he described when interviewed:

> It was so frustrating in Greece....Speed boats go up and down the place and kill turtles, and one day I rescued one of these animals....It had been sliced by a speed boat, and when I was bringing it back it was just having its last shakes before it died. We couldn't have done anything for it, and I was with these guys who [suggested] direct action....There were fifty boats and they were all illegal [so] we spent the whole night 'altering' boats' engines, and for the next week no boat was going up and around the bay.

'Lorenzo' joined EF! after arriving in the UK as a student. Some of the second cohort were bitterly critical of existing groups. 'Daniel' described FoE as 'politically conservative and liable to bottle-out of any situation...they had a camp at Twyford Down and were told to leave, and left. It is the sort of organisation you can still be a Conservative Party member and join!' In dramatic contrast to 'Daniel', Freeman worked for FoE as a paid officer, and continued to do so after joining EF! (UK)'s Reclaim the Streets' campaign in 1992. She was frustrated by their initial failure to campaign against the car:

> I kept saying to FOE, 'Why don't we do a car campaign?', and they would say, 'You do need a car in some areas, a lot of our members have cars, [we must] not offend people.'...I remember the day I walked through Cooltan thinking, 'This is going to change my life'. [In] 1992 [I] heard about it, painted cycle lanes, did the Motorshow and it was an incredible group of people.

Cooltan was the name of a former suntan-lotion factory which Collins opened as a squatted centre for green activity in Brixton, south London. Unlike Collins or Freeman, others had come from a left or an anarchist milieu, and were new to green politics but found EF! (UK) and the wider anti-roads movement attractive.

> I was quite interested in reading stuff about the Solsbury Hill campaign, and that drew me towards [green politics]. Then I got involved with the Bristol Criminal Justice Act and did a few actions and met Earth First! people and read the things they put out. I liked what I read. I read *Do or Die!* about 1994....I thought this was bloody good, people getting involved and doing direct action, because when I did my first direct action that was good fun. At Solsbury Hill, apart from the security guards being a bit dodgy, it was fun jumping on cranes and getting dragged out of the site only to run back and be dragged out again. I have always been a person who has believed compassionately [*sic*] about something, and I want to get out there and do something.

> I was doing student union stuff, grant cuts....I have always been a political person, really. I quit the SWP and thought Earth First! are a bit better. They are not trying to recruit people, push their own opinions on to people all the time. It is more of a coalition of people, different views, and that's about it.
>
> ('Mark' in interview)

Such activists from leftish milieux contrasted the enthusiasm of what they perceived as a new politics with their early forms of participation. In interview 'Green' noted, simply: 'By the early 1990s everybody was so apathetic and demoralised there wasn't anything going on.' As the next chapter will argue, such demoralisation may be linked to wider shifts in perceived political opportunities in the UK.

Green culture

Individuals from both cohorts increasingly took on dual identities. They became *activists* and they became *greens*. Most of the first cohort perceived themselves to be both 'greens' and 'activists' prior to their EF! (UK)/anti-roads participation. Of the second cohort, some, at least, held an activist identity, but were not initially 'green', often entering a phase of intense cultural and political transformation within EF! (UK):

> It took a good couple of months to get my head round it....Before my involvement I think I saw the green strand as quite marginalised and middle-class, somehow quite softy....I stayed at the M11...got involved with Wanstonia, and then kind of had this part-time existence in the campaign. Because it is quite a mad culture....It took me quite a long time to work out what on earth it was all about, which I am still doing...and it became that kind of community thing...recognising people and that solidarity with the people. So my focus shifted to east London and I squatted...[and] was integrated into the campaign as a whole.
>
> ('Green' in interview)

Such an account reflects McAdam's view (1986: 72) that strong participation in intense movement activity may have a 'profound socialising effect on the individuals concerned'. In this sense direct action may be like a ritual that eases the often traumatic and anxiety-inducing passage from one identity to another.

Rebecca Lush became active at Twyford via counter-cultural networks: 'I was more socialist before I was green....The green thing was very linked to the hippie thing, so I leaned to the hippie identity and started going to these free festivals...[where] little bits about Twyford filtered through all summer.' After a series of adventures, including being arrested as part of a new age traveller convoy and visiting Stonehenge, Lush became a full-time camper at Twyford Down.

Becoming an activist demanded that individuals learn a series of practices loaded with both material and symbolic importance for 'green' and 'activist' identity:

> [As] the summer went on [I] learnt how to hitch, how to bunk trains,...all the little skills you pick up, that you should be taught at University, of direct action, just learning the ropes...how to DTP a leaflet, how to, you know, all that sort of stuff and, yeah, [I] slowly but surely became what I considered quite fully skilled in the art of being an Earth First!er. In terms of inter-group stuff, there were a few dreadlocks flying around, but I was still into beer-drinking, playing football and sometimes political correctness....I have probably improved now, improved is a very loaded word but...I was suddenly into something that was very different from my previous twenty years or so.
>
> (Styles in interview)

Availability

As predicted by McAdam's account (1986: 85) of biographical availability, activists from both cohorts had few family or work commitments to restrict their participation. The overwhelming majority were either university students (8 plus one further-education student) or unemployed (14) when they first became involved in EF! (UK). A further three were self-employed, a category which may provide some personal time flexibility.

While increased availability helped to increase participation, as an activist identity was embraced work or study commitments might be consciously reduced to enable more intensive commitment. Geffen noted how self-employment was replaced by activist involvement:

> Twyford was terribly inspiring, the non-violence thing, the dedication and just meeting brilliant people was all very exciting, and that increasingly replaced the LCC thing in my spare time, which there was quite a bit of, having dropped out of full time work. I was working as a record producer for a small classical music company near Oxleas Wood....I left that hoping to find freelance work....So through all this time the campaigning was filling what was an unwanted vacuum, but as time went on I was increasingly not looking back.
>
> (Geffen in interview)

Lush, as noted earlier, dropped out of university to live as a Donga. Freeman, one of just two waged interviewees in the sample, had seen biographical availability lead to an activist role in the 1970s when she became a volunteer for FoE. Ironically, in the 1990s her position as an FoE employee sometimes conflicted with her role as an anti-roads activist.

Few activists had children under the age of 16 or other caring commitments. None mentioned a caring role that might have restricted activism. Nonetheless intensive activism could present a challenge to personal relationships. Collins noted: 'I started and stuck it out for a year-and-a-half, but couldn't manage college and Green Party and Cooltan and home-life and love-life, so I knocked the college on the head.' Almost all the activists interviewed were young. Yet media accounts and my own observations suggest that the requisite biographical availability has led also to the participation in environmental direct action of older and retired people, who equally have relatively few employment or family caring commitments.

Merrick notes, conversely, how family commitments cemented otherwise sympathetic individuals into opposing the anti-roads movement:

> Earlier today, one told me how he grew up not far away, in Didcot, that he knows this area and loves it and knows that the road is wrong, that it's just political games. That if he was young and single, he wouldn't just not be here as a guard, he'd be here with us. But he's got two kids under 5 and it's the first work he's had in three years.
>
> (Merrick 1996: 32)

Being green

EF! (UK) was initiated by existing green activists who sought to create a network that would be more decentralist and committed to environmental direct action of a confrontational kind than was the case in the broader green movement. Their accounts closely parallel those of EF! (US) founding activists and other green political entrepreneurs who sought to create 'radical' new environmental movements (Foreman 1991; Lee 1995: 25–37; Manes 1990: 66–70; Scarce 1990: 58–62; Zakin 1993: 132–4). Typically, Rucht notes that the 'determining factors for the founding of Earth First! [US] were a growing dissatisfaction with the structure, practices and organization of the large, conventional and resource rich environmental organizations in the United States' (1995: 74). Jenkins (1983: 531) argues that new groups, in protest movements as a whole, are generally created by disillusioned activists who split away from existing groups.

Founding EF! (UK) activists felt that the strategies of existing groups were increasingly ineffective as well as eroding the essential values of participation and empowerment. Beyond this common factor of dissatisfaction with existing green groups, routes into direct action were varied. The meaning of activist involvement was different for the individual interviewees, often carrying with it an intense commitment that cannot easily be captured in words. Torrance observed:

> I was really living on this kind of passion, this...real kind of excitement. Things were really happening, and I just moved out of my flat in Hastings, dropped out of college....Jake as well had dropped out of college. We both dropped out of our exams to get EF! going.

Such enthusiasm can be dismissed, in Torrance's own words, as 'raw naivety', and EF! (UK) might be conceived of as a youth movement motivated by a rejection of parental and social authority. Secrett, the Director of FoE, referring to these founding activists, argued:

> One of the things that pissed me off when I came back into FoE – and now I am caricaturing to make the point – [was] these 20-somethings who told me I had it all wrong and theirs was the only way of contributing. 'Who the fuck are you? I can guarantee that in five years time 80 per cent of you will be working in banks and solicitors.'...I am going back to [that] Twyford Down time [when relations between FoE and the direct action movement were particularly poor].
>
> (Secrett in interview)

Secrett's bitter comments suggest that EF! (UK) can be depicted in terms of a recurrent rebellion of youth against authority, a rebellion that has often incorporated a green dimension and support for direct action tactics. The wider anti-roads movement has been marked by the participation of numerous young activists (Merrick 1996). Parallels to both EF! (UK) and the youthful counter-culturalists such as the Dongas tribe may be seen in the original Woodcraft Folk, the European *Wandervogel*, or the late 1960s counter-culture of young people in the USA (Paul 1951: 55–63; Reich 1973; Roszak 1972).

One founding activist, introducing the notion of a specifically male approach to identity, argued:

> Especially if it is a young male confronting an older male in an established NGO, then we have this kind of father–son generational difference as well, which for a woman of any age looking on is really boring. Its actually got nothing to do with FoE or Greenpeace or whatever. It's just an angry young person that's not really accepted by the sort of conventional NGO scene.
>
> (Zelter in interview)

Such a viewpoint does not explain why young radicals did not join the existing GSN, *Green Anarchist* or militant groups outside the green movement; nor does it explain why the anti-roads protest revived so strongly during the 1990s. Noble's account of rebellion against parental authority indicates the ambivalence of such a 'move'. While rebelling, she also felt that her green childhood put her on a pathway to an adult green identity. Finally, Zelter's suggestion does not account for her own attraction to EF! (UK) and that experienced by other women activists, particularly those who participated at Greenham.

Another possible influence is the existence within the family of what can be termed an activist culture. Bagguley (1992) argues that social movements are likely to contain individuals resourced with movement 'skills', confident at running meetings, writing press releases, designing posters. Such individuals will be particularly effective and willing to carry out activities such as media work, event organisation, legal defence, etc. Several interviewees from both cohorts appear to have come from households where political and other forms of community involvement were common, and whose parents were variously local councillors, shop-stewards or others who played an important organisational role in the workplace or the community – Noble's fundamentalist Christian father, for instance. Anti-road activists outside of EF! (UK) often included other highly politicised local organisers, such as Tommy Sheridan, a leader of Militant Labour in Glasgow, and former Conservative local councillors Barbara Bryant and David Croker.

As we have seen EF! (US) initial activists included several individuals who had been highly active organisers in such conservation SMOs as the Sierra Club. Judi Bari had been a union activist and Foreman had been a Republican organiser.

In summary, activist participation in EF! (UK), as the social-movement literature suggests, is rooted in a process combining previous activism, political and friendship networks, personal political sympathy and biographical availability. These factors cannot, at least for the EF! (UK) interviewees, be aligned, as they are in some accounts of activist trajectories, to form consecutive steps leading to intensive activism. Some individuals perceived their prior sympathies to have influenced activism, while for others network participation was seen as crucial to the transformation of their attitudes.

Consideration of individual availability illustrates this point. EF! (UK) activity is very demanding of time, far more than participation in pressure groups or the Green Party, because of the need for involvement in direct action that may include continuous participation, such as residing at an anti-road camp. Secrett observed: 'It is a difficult thing living rough. One has deep-rooted admiration for people at protest camps over months, [in] terrible conditions, freezing cold pouring rain.' Equally, for the founding activists the act of creating a new movement is always likely to be demanding of time and personal energy. The sample who fell into the categories of student, unemployed or self-employed mirror the typology that researchers have identified as likely to be activists due to their flexible time commitments. Similarly, the activists had few or no strong family commitments or dependants at the time. Yet while interviewees were thus 'biographically available', such availability was often increased as a conscious political choice. Lush and Torrance rejected student life for full-time activism, while Geffen moved from self-employment to intensive anti-roads campaigning.

Of the second cohort of activists (those who had joined EF! (UK) after April 1992), several had not previously been members of or activists in the green movement, and one saw attitudinal sympathy as a product rather than a prerequisite of EF! (UK) participation. For these individuals, participation cannot be explained in terms of prior green identity or dissatisfaction with green ideology. Thus we need to ask why EF! (UK) succeeded in attracting the support it did, as well as that of other allies. Greenpeace London (the anarchistic green group which is not to be confused with Greenpeace International) was founded in 1970 and *Green Anarchist* in 1984, while Laughton had attempted, unsuccessfully, to launch EF! (UK) in 1987. Why did these attempts to create direct action-oriented, libertarian, green networks fail to mobilise activists, in contrast to EF! (UK)'s success in 1991–2? Equally, why did anti-road protests, following more than a decade of decline after Tyme's 1970s actions, take off during the 1990s? To understand more, the micro-sociology of personal identity formation must be linked to larger social and political change, a topic treated in the next chapter.

6 Political opportunities and direct action

> But maybe it is Mrs Thatcher who should be thanked most because many of our activists were unemployed, early retirees or redundant and were able to give all of their time to fighting the campaign.
>
> (Charles Elstone quoted in Stewart *et al.* 1995: 22)

> Blott knew the British too well to suppose they would do anything to endanger life. And yet without endangering life, and Blott's life in particular, there would be no way of building the motorway on through the Park and Handyman Hall. The Lodge, now Festung Blott, stood directly in the path of the motorway. On either side the cliffs rose steeply. Before anything could be done the Lodge would have to be demolished and since Blott was encased within it, demolishing the arch would mean demolishing him.
>
> (Sharpe 1975: 205)

Introduction

In November 1990, five months before EF! (UK)'s first public action, Mrs Thatcher was forced to resign after more then eleven years as Prime Minister. Cabinet divisions over European Monetary Union acted as the immediate trigger for her removal. Widespread public opposition to the introduction of the Poll Tax, including mass non-payment and rioting, contributed to her demise (Watkins 1991: 27–8). 'Thatcherism', whether conceptualised as idiosyncratic personal belief, distinctive ideology, part of a global right-wing shift or as some other phenomenon, influenced the green movement, albeit in varied and contradictory ways (McCormick 1991; Robinson 1992; Rüdig 1993). Mrs Thatcher, seen initially as implacably hostile to the green movement, helped to make environmental concerns, particularly issues such as global warming and ozone depletion, 'respectable' during the late 1980s (see Chapter 2 and McCormick 1991). Despite this belated and partial acknowledgement of the environment's importance, many believed that her departure would signal a political realignment and progressive change. In the event, she was replaced by John Major who, to the surprise of some, was then re-elected as Prime Minister in April 1992 and continued in government with a largely free-market agenda until his 1997 defeat (Hogg and Hill 1995: 1). In short, in 1991–2, EF! (UK) and the anti-roads movement mobilised in a highly specific political context. To what extent, if at all, did the exigencies of this context, allied to 'macro' factors at the national level, favour, retard or shape such mobilisation?

Some academics have argued that mobilisation can be explained by examining the specific political context or the political opportunity structure (POS). Eisinger introduced the POS concept, stating: 'The manner in which individuals and groups in the political system behave…is not simply a function of the resources they command but of the openings, weak spots, barriers and resources of the political system itself' (1973: 12). He suggested that protest movements were unlikely to occur in either 'open' systems, where demands could be met without agitation, or 'closed' ones where severe repression discourages protest. Movements were most likely to become active where a mix of 'open' and 'closed' features occurred.

Tarrow (1994: 17) argues that 'the "when" of social movement mobilization – when political opportunities are opening up – goes a long way towards explaining its "why" '. Drawing on RMT, he claims that a shift in the POS from a relatively 'closed' position to one which is more 'open' creates the possibility for collective action by lowering the 'costs' involved for potential participants. The 'opening up of access to participation, shifts in ruling alignments, the availability of influential allies, and cleavages within…elites' are all relevant (Tarrow 1994: 86). Mix's comment – that both EF! (UK) and FoE became involved with the Twyford Down campaign because they felt the expected election of a Labour government in 1992 would give them victory – accords with Tarrow's predictions.

Kitschelt (1986), studying anti-nuclear power movements, identified three factors that 'further or restrain the capacity of social movements to engage in protest': cultural norms and traditions of strategy; the policy process; and, finally, the existence and approach of other movements. He placed particular importance on the policy process in terms of its openness to movement demands and its implementation ability. In open systems which have a strong ability to implement policy, movements are predicted to pursue 'assimilative strategies…[such as] lobbying, petitioning…referendum campaigns and partisan involvement in electoral contests' (Kitschelt 1986: 67). In closed systems with a weak ability to implement agreed policy, movements are more likely to use '[c]onfrontational strategies…[such as] public demonstrations and acts of civil disobedience, exemplified by occupations of nuclear plant sites and access roads'. Such confrontation results because movements feel not only that access to the policy-making process using conventional political means is blocked but that weak implementation procedures may make disruption using direct action effective.

In terms of Kitschelt's definition of the likely features of a relatively closed policy process, the UK might be seen as having long been a candidate for the emergence of confrontational movements using direct action strategies. The UK, in contrast to most Continental European states, has a non-proportional electoral system, which has blocked green parliamentary representation (Rootes 1995: 68). The executive, in the form of the UK Prime Minister and cabinet, controls parliament, and the period (1979–90) saw policies developed under instruction from cabinets dominated by a strong leader who relied to an increasing extent on external think-tanks. Equally, large parliamentary majorities have insulated government from public pressure between elections. In fact, 'Great Britain provides the classic example...of a single-party government, where the governing party is highly disciplined. Such governments generally have a strong capacity to act' (Kriesi *et al.* 1995: 30).

It has been argued also that, in contrast to Continental Western Europe, the UK lacks 'new social movement' activity and left-libertarian politics, and that the green movement, such as it is, has largely rejected unconventional political strategies. A number of arguments have been used to account for such 'British exceptionalism' (Rootes 1992; Rüdig and Lowe 1986). One might assume from Kitschelt's account that given the strength of the UK system, as well as its disposition of relative closure, confrontational strategies would be seen as simply ineffective and lifestyle politics would be more attractive than protest. Advocates of 'British exceptionalism' suggest that green demands have been integrated with conventional forms of political activity. For example, the Labour movement has been seen as capturing and subsuming parts of the family of green and other social movements (Eyerman and Jamison 1991: 37). In fact, British exceptionalism seems to be more limited than was once thought: road protests in the 1970s utilised direct action, and during the 1980s hundreds of thousands of activists marched for peace. Militant animal liberationists have long invaded laboratories, released rats, cats and chickens, persecuted animal abusers and even burned down department stores that were selling fur. Yet EF! (UK) and the wider anti-roads movement are indicative of a dramatic increase in direct action, perhaps influenced by political factors.

Activist accounts

Activists touched on likely sources of political influence at different stages during the interviews. The opening invitation to discuss personal participation often produced responses that touched on macro factors, while the more specific issues were dealt with at a later stage. One question asked why activists felt that EF! had emerged in the UK and another whether they felt that Thatcherism had influenced their own participation. A fertile line of questioning looked at activists' perceptions of the UK POS in contrast to that in other European states. This follows the common POS procedure of cross-country comparison, so as to identify the political features that most directly concern activists.

Thatcherism and Earth First!

Relatively few activists touched on Thatcherism in the first part of the interview, where an introductory open question invited them to describe the process by which they came to participate in EF! (UK). Even when questioned directly, many were bemused – many, after all, had become politically active and/or environmentally concerned only during the last months of the final Thatcher administration. Others saw Thatcherism as so widely invasive a phenomenon that they could not account for any specific influence.

Asked whether Thatcher had influenced his environmental strategies, Begg replied:

> I haven't the faintest idea. For me, Mrs Thatcher is not sort of one Prime Minister amongst many. She is *the* Prime Minister because she is really the only experience of a Prime Minister I have had, unless you count John Major, and really I don't....Mrs Thatcher isn't just a politician, Mrs Thatcher *is* politics. She's defined the entire political culture I am operating [within]. It is impossible to say to what extent I am responding to Mrs Thatcher.

Often only simple hostility to her policies was evident. Her legitimising of environmental ideas, together with a more enduring hostility on her part to green politics, was noted by some, Collins, for example, said:

> Her, you know, conversion to green thinking...gave a sort of respectability in a funny sort of a way...to a lot of the green stuff....It was very short-lived, but at the time it did sort of make it nice to be green....And then, on the other way, her policies made one very, very annoyed on every level...socially, environmentally, whatever, so I think that was...a spur as well. So, yeah, maybe she helped out in both ways.

Two activists, Begg and Marshall, were unusually reflective on this issue. Both were social science graduates who as students might, therefore, be expected to have attended more closely than other activists to defining the Thatcher phenomenon. Marshall, of course, had the added advantage of having been a member of his Conservative Students Association, as well as having attended meetings of the Revolutionary Communist Party while at university!

Political closure and the illusion of opening

Founding activists felt, despite the excitement of the late 1980s, that even modest environmental demands had dropped from the political agenda by 1991. 'Mary' suggested that

> people in the Green Party...were feeling quite disillusioned...[there] had been a lot of big push...of you know, let's go environment-friendly and have lead-free petrol and recycled paper....There was a lot of frustration about that when the big issues were not being addressed at all, i.e. let's bugger lead-free petrol, let's cut down on car use, let's get rid of cars....Again, there are a lot of young people feeling quite angry, and they couldn't get involved in it any other way. This was the only way for them to get involved in it, and I think Jake and Jason picked up on that...they got the timing right. They were the right people in the right place.

Other greens recognised similar political features during this period. Typically, in May 1991, *Green Line* argued that the 'first great wave of green politics' had begun in 1988, had peaked with the 1989 Euro-elections, and now was over, and that the movement had been 'effectively neutralised' (Andrews 1991: 4–5).

The Greenpeace UK Programme Director noted:

> The 'green wave' hit Britain in the late 1980s. Seals died in the North Sea in 1988, and Mrs Thatcher 'went green' at the Royal Society. The Greens won 15 [per cent] at the 1989 European Elections and the Government published its White Paper *This Common Inheritance* in 1990. During these years membership of environmental organizations grew exponentially.
>
> Then suddenly it was over. At the 1992 Earth Summit the television images of the struggle did not match the aspirations of the audience. The environmental messages of Rio fell like ashes in a shower of disappointment....Within most environment groups the years 1989–92 were marked by inward self-examination, training, management changes and restructuring to cope with the results of relatively massive growth, swiftly followed by the onset of recession.
>
> (Rose 1993: 288–9)

Thus it can be argued that 'if 1989/90 was a period of green euphoria, then 1990/91 could be classified as a time of green fatigue' (Rüdig 1993: 1).

Organisational mediation

Existing environmental pressure groups and the Green Party, responsive to shifts in the policy-making process, provided an important link between political opportunities and EF! (UK)/the anti-roads movement. The apparent openness of the POS to environmental demands in the late 1980s encouraged many groups to moderate strategy, to enhance policy links to government, to resist activist pressures for more confrontational stances and to 'streamline' internal organisation. An apparent opening in the POS promoted political 'realism'. Yet when the POS remained closed, such realism appeared increasingly inappropriate: one is tempted to say unrealistic. Continued closure increased support for direct action, weakening 'realist' positions and easing the emergence of EF! (UK) and anti-road campaigns.

For example, within the Green Party, a politically 'realist' faction achieved control at the 1991 Autumn Conference, arguing that the proximity of ecological crisis meant that the Party should aim for government in the early years of the 21st century. To gain such 'an electoral mandate' demanded a restructuring of internal organisation and a commitment to 'accept the need for leadership as a function' and 'work within the existing system' (Green 2000 1990). An exclusively electoral role for the Party was written into the constitution and put into practice by the new executive (Andrews 1991; Costello 1992; Wall 1991a). Extra-parliamentary politics were frowned upon.

The huge 1989 vote suggested that the POS was open to the participation of an electoral Green Party within the UK parliamentary system, despite the absence of proportional representation. In turn, the exclusively parliamentary strategy assumed by Green 2000, with the target of eventually electing a Green Party government, gained legitimacy among members.

The wave of recruits for the Party in 1989 was influenced to a large extent by media representations of the environmental and publicity campaigns by Greenpeace and FoE. These members measured success largely in terms of the creation of media images and often dismissed the decentralist element of green politics (Evans 1993). Identifying with the environmental pressure groups, these members strongly supported Green 2000, providing the political 'realists' with a strong constituency of supporters committed to introducing constitutional change.

Green 2000, in its turn, argued that the declining poll ratings and membership after 1990 were products of the Party's constitution which was viewed as inefficient, unrepresentative and failing to provide the kind of leadership necessary for renewal of strength (Lambert 1991).

Their 'fundamentalist' opponents, myself included, felt that, given the organisational and ideological hegemony of Green 2000, the possibility of reversing such changes was remote. Many withdrew from political activity, others concentrated on local Party work, while a number looked for new organisational forms that better reflected their decentralist concerns. EF! (UK) was one such alternative.

In 1992 the General Election produced a low poll for the Green Party. By the end of the year Parkin and other key members of Green 2000 resigned from the Party Executive, and the decentralists regained influence. One commentator noted that

> the turmoil within the Green Party is simply one symptom of the wider crisis. Other signs include the violent denunciation of market environmentalism...and the haemorrhage from FoE of local members who are frustrated

> by the restrictions placed on them by the leadership and are attracted by the more confrontational direct approach of anarchist influenced groups.
>
> (Dickson 1992: 10–11)

Equally, although Greenpeace International, in contrast to the libertarian GL, had long rejected participatory internal structures, it is clear that the political opportunities both in the UK and internationally encouraged it to embrace more assimilative strategies between 1988 and 1992. Greenpeace believed that greater political awareness of international issues such as global warming reduced the need for confrontational tactics. Fostering co-operation with government and industry became more important. Wilkinson, who was a founder of the organisation in the UK, in 1991 accused it of 'isolationism and loss of direction' (Wilkinson and Schofield 1994: 127). He stressed that such a lack of dynamism was part of the general green-movement malaise believed to have resulted from shifting political opportunities.

> On my return from New Zealand, it had taken me a few weeks to get a handle on what the green movement was up to in the UK....The scene was flat. Nothing appeared to be going on. It was most depressing. Mrs Thatcher had pulled the rug from beneath the feet of the greens by her 'we are all green now' speech to the UN, and everyone was staring at each other, saying 'What now?'
>
> (Wilkinson and Schofield 1994: 128)

FoE also moved towards more assimilative strategies, as well as greater centralisation in the period 1987–92 (Lamb 1996; McCormick 1991: 117; Weston 1989). Porritt, its Director between 1984 and 1990, noted how FoE had shifted 'since 1987, from campaigning and working *against* industry to working increasingly *with* industry', observing:

> FoE spent a lot of time beating at doors because people didn't want to let us in. We had to work hard to get the attention of ministers....Now we just have to nudge the door, it opens wide, and we fall flat on our faces, not really knowing what to do when we get there.
>
> (Porritt in McCormick 1991: 117)

McCormick further argues that practical advice had replaced 'criticism and confrontation' (1991: 118). The FoE Annual Report of 1989/90 confirmed that the organisation had 'developed from opposition to proposition', with a resulting adaptation of 'campaigning tools' (Huey 1990: 3).

There is also evidence that a more assimilative strategy was linked to a less libertarian organisational form. Although former trade union leader David Gee had replaced Jonathon Porritt as head of FoE, many saw the late Andrew Lees, who became campaigns co-ordinator in 1989, as the dominant figure in the organisation (Marshall in interview). He was particularly hostile to grassroots activists, such as those individuals who came to form EF! (UK), and was often mentioned in negative terms by interviewees. Describing his approach to FoE on behalf of the London Rainforest Action Group, Marshall noted:

> I remember going around to Friends of the Earth and saying: 'I wonder whether I could look at your files?', and they just completely freaked out....[That] they would actively contact their local groups, autonomous local groups, and tell them not to get involved in our stuff was disheartening and actually was a declaration of war. And this was all very much coming from Andrew Lees at this time, this was June 1991.
>
> (Marshall in interview)

Sheila Freeman, an FoE employee who became an EF! (UK) activist in 1993, made a similar point:

> I went on the M11 campaign that FoE also didn't get involved with at all. It was the same thing again [as at Twyford]. They thought it was too anarchic, [that] it [had] too much direct action, not enough public sympathy behind it, and it was down to Andrew Lees who was the campaigns' co-ordinator. He was an interesting campaigner. He just lived and breathed the environment, it was his life really, and on his own he probably would have been into direct action. But he didn't think it was what an organisation like FoE should be involved in. He was worried we would lose our members [and] we would lose our funding.
>
> (Freeman in interview)

A history of FoE produced in association with the organisation also noted criticisms of FoE from local activists in the period between 1989 and 1992 (Lamb 1996). One prominent FoE member claimed that Lees 'operated a kind of medieval patronage system. If he thought you [a local FoE group] were good, he'd give you anything', while less-favoured local groups were 'left out in the cold' (Lamb 1996: 166).

Such accounts situate Lees' approach in a wider political context:

> The momentum of FoE's campaigns seemed to some onlookers to slacken in this unwonted atmosphere of official approval: when the Government set a new challenge for local authorities to recycle at least 25 per cent of municipal wastes, for example, FoE and its local groups found themselves in instant demand as advisers. Even so, discontent among local groups was surfacing

> once more. Some felt put out by the national organisation's income-generating activities, which they saw as siphoning off local support and courting the armchair donor rather than the local activist. Nor did they see benefits coming their way from the increased spending-power the fund-raising bestowed. Others felt the organisation was becoming ineffective as an agent of change in relation to government and industry.
>
> Still others felt excluded from the campaigns side of things.
>
> (Lamb 1996: 166)

Frustration with FoE particularly influenced EF! (UK) activists, in contrast to their disillusionment with Greenpeace and the Green Party. Greenpeace shifted towards a more assimilative strategy but had *never* encouraged grassroot-activist involvement. Thus while EF! (UK) activists had an almost universally negative view of Greenpeace's attitude to participation, there is no evidence that the organisation's anti-libertarian approach significantly changed in the late 1980s. Many EF! (UK) activists who were members of the Green Party, including Begg, Collins and Noble, maintained their membership of the Party after the Green 2000 victory. While Green 2000 gained national control, local parties were practically independent, and one activist went so far as to claim: 'We had a so-called Green Party group in Littlehampton that also went under the guise of Earth First! It was west Sussex Green Party and Arun Earth First!' (Noble in interview).

This continuity of membership is significant for two reasons: activists felt that they could retain membership and participate in EF! (UK) direct action; and, more significantly, they were not disciplined by the central Party for doing so, even when they took part in such events as Party representatives. The most likely explanation was that the Green Party executive was simply incapable of tracking such activists and transmitting discipline to local branches. Greenpeace in contrast had a strong transmission belt whereby officers were dispatched to intervene in local branch organisation; equally, FoE took a firm stand and discouraged members co-operating with EF! (UK) until 1994.

Alternative movement activity

Shifting priorities within the green movement and between other movements also aided EF! (UK)'s emergence:

> In the late 1970s and early 1980s, the peace movement (in Britain, at least) tended to overshadow the green movement, and the liberals and radical liberals who might otherwise be campaigning against nuclear power (and other environmental issues) were campaigning against nuclear weapons.

> Since the peace movement went into decline, the environment movement has grown correspondingly.
>
> (Anon. 1994: 13)

A number of founding activists, some of whom had been highly active in the peace movement, broadly endorsed this suggestion (as indicated by remarks made by Torrance and 'Mary' in interview). With the thawing of the Cold War, peace clearly fell down the agenda for greens, and other issues threatening planetary survival took its place. Other activists had transferred their energies from animal liberation campaigning or participation in other elements of the broad green movement family to EF! (UK).

As noted in Chapter 5, activists from the second cohort of interviewees were found increasingly to have participated in other movement families. Activists who at one time might have supported the Labour movement, participating in the Labour Party, far-left groups or trade unions, seemed less likely to do so by the early 1990s. One activist argued:

> The working-class lost pretty much all of its power in the 1980s, so people didn't get involved in the Marxist and anarchist workerist groups.... What is the point of joining a group with no power? If you are going to move something, you need a lever. Power is crucial...[and loss of working-class power] led young militants to reassess and look for new ways of doing things.
>
> (Anon. in interview)

Of the second cohort of interviewees, several activists with no previous green involvement left the Labour movement because of their rising environmental concern and a common perception that the green movement had potential as a vehicle for wider and more radical political activity (see Chapter 5). In this sense activists moved, as in the title of Bahro's book, *From Red to Green* (1984).

Although links between the green movement and the Labour movement have been limited since the decline, around 1900, of the earlier form of green politics, such crossovers continued into the 1980s. Rootes (1992) has noted the connections, in the early 1980s, between the Labour leadership and the Campaign for Nuclear Disarmament (CND) and argues that the Labour Party was 'a broad church encompassing a variety of tendencies...permeable to young radicals in a way that, say, the German SPD [Social Democratic Party] was not' (Rootes 1992: 184). Although agreeing that the national Labour Party largely rejected green concerns, Rootes argues that the populist style of socialism embraced by the more left-wing Labour local councils, such as Ken Livingstone's GLC, promoted 'environmentalist initiatives'.

Yet the Labour Party ceased its support for CND in the late 1980s, and local socialism was marginalised both by Thatcher's local government reforms and by the Party leadership, while generally the Party became less sympathetic to social movement demands (Heffernan and Marquesse 1992). Political realism intensified with the election of Tony Blair as Labour Party leader in 1994. On the eve of the 1996 Labour Party Conference, sacked docks workers held a street party and a direct action protest with Reclaim the Streets and other EF! (UK) groups (Anon. in interview; see Pilger 1996a). The dockers approached RTS, after the latter had attempted to occupy the offices of London Transport in support of striking tube workers, who were seen as possible allies and a source of new repertoires of action.

Repression and facilitation

Repressive action has been a feature of the UK state's responses to animal liberation, anti-nuclear power and peace-movement activism, plus that of the green counter-cultures; facilitation, while rarer, has also occurred (Goodwin 1995: 22–3). Repression has been defined as 'any action by another group which raises the...cost of collective action', and facilitation as any which lowers that cost (Tilly 1978: 100). Both effects may interlink as complementary elements of a strategy to discourage radicals by making their actions more difficult, while simultaneously assisting movement moderates to mobilise in less-threatening ways (Kriesi *et al.* 1995: 124; Marx 1979: 106). Thus, facilitation may lead to 'increased levels of movement mobilization' but also restrict mobilisation to 'moderate' repertoires (Kriesi *et al.* 1995: 38). In contrast, low levels of either repression or facilitation may make direct action mobilisation more likely (Kriesi *et al.* 1995: 39).

Although 'governments specialize in the control of mobilization and collective action: police for crowd control, troops to back them, spies and informers for infiltration' (Tilly 1978: 101), repression involving diverse state (and non-state) agencies may lead to fractured and contradictory agendas. 'Turf-wars' between different agencies may occur and state agents may promote militancy as a means of increasing their own access to scant state resources (O'Hara 1994: 94). State agencies may also help construct counter-movements, which oppose the aims of green or other radical activists; equally, corporations may employ their own agents of repression (Marx 1979; Tilly 1978: 101). With the 1990s expansion of private security services in the UK, boundaries between counter-movements, state agencies and capital have become increasingly blurred (Gallagher 1995: 24; South 1988: 9).

The 'costs' to activists, in terms of injury, arrest and other counter-measures, were commonly noted by interviewees. 'Repression' was exercised in several ways. First, the uniformed police officers who made arrests occasionally used violence against activists. The Criminal Justice Act of 1994 criminalised many acts of NVDA and extended police powers against activists (Collin and Godfrey

1997: 227). Such an increase in the powers available to the police seems to be part of a long-term trend in disciplinary control, which accelerated during the mid-1990s. Second, road contractors and other corporate bodies threatened by EF! (UK) and other environmental direct action campaigners employed increasing numbers of private security guards. In direct response to anti-road activists, the DoT advised private contractors to add security costs to tender documents (Smulian 1994: 1). Guards seem to have used increasingly violent tactics since December 1992 (Crossley 1995: 14; Gallagher 1995: 23). Third, intelligence-gathering, covert and overt, has occurred, as illustrated by the activities of the Forward Intelligence Team (see p. 127).

Acts of violence were commonly noted by interviewees. Thus Styles observed:

> My first arrest was at Abbey Pond. So, we had been in the pond for an hour-and-a-half, we had finished all our sherry and then we came out and we were arrested for basically walking down the road. So I thought this was highly unjust and…I was booting the top of the police van with my steel-toecapped Dockers and going absolutely mad. The Manchester police didn't take a shine to me either, and I basically got assaulted. I got beaten around the knees and roughed about on the way to the cells.

James noted: 'I was kicked in the testicles by a policeman at Euphoria [an anti-road squat in east London], but that was very run-of-the-mill stuff for that campaign. I knew people who were getting assaulted much worse.' In December 1992, apparently for the first time, private security guards were used against an anti-road camp. Lush graphically described the violence used at Twyford Down:

> They took three blokes behind the machinery and really beat them, gave them a systematic beating, as the police surgeon said when he saw them….I saw them going to police-vans with blood streaming down their faces, and later Roe was strangled until she was unconscious and I was strangled [by police] and kicked in the ribs a couple of times by a copper.

The details of her story were reflected in a significant number of journalistic and independent accounts.

O'Hara has argued that the end of the Cold War, the decline of the far-left and the ceasefire in Northern Ireland created a crisis for UK military intelligence in the early 1990s. Protest-policing provided one alternative, though much-reduced, role that might be used to legitimise the continued existence of intelligence agencies during a thin period. O'Hara outlines two cases (1993a and 1993b; and 1994: 88–90) of apparent intervention in EF! (UK) by military intelligence. In one, Tim Hepple, an advocate of militant EF! (UK) repertoires, apparently acted as an *agent provocateur* (see Chapter 1). Hepple publicly acknowledged his role as an infiltrator of the far-right British National Party, and confessed to being an *agent provocateur* for the green movement. Hepple has written a fascinating account (1993) of his career as an entrepreneur of violence, in which he recounts his trajectory from far-right activist to far-right infiltrator and militant green. In the second case identified by O'Hara (1994), attempts were made apparently to establish a false support group for environmental-activist prisoners.

A uniformed Forward Intelligence Team (FIT), drawn from the police, was established to monitor key green activists in the mid-1990s. FIT activities seem to reflect police fear of loss of control resulting from the 'temporary autonomous zones' created by mass direct action at street parties and other anti-road protests. For example, the Metropolitan Police reacted strongly to the 1996 occupation of the M41, my own encounter with the FIT (see p. 2) coming two days after the event and the raid on the RTS–EF! offices in London following quickly. Increasingly, private investigators were employed by the DoT to gather information on anti-road activists. Observation of activists was carried out most commonly as a means of facilitating the serving of legal injunctions. Direct intimidation may also have been a motive.

Such intelligence gathering was noted by activists:

> For a time I was followed by the police. I didn't pick up on this at all, until basically being arrested in 1995 on charges of incitement to criminal damage.... A few weeks before I was driving in my car and there was this other car behind and then, as I went past this side road, this other car pulled out, getting in between me and the car that was some way behind. The car that was some way behind then accelerated, overtook the car that had got in between us, and then pulled back behind me. I thought: 'This is a bit strange. It seems as if this car is following me.' But I didn't want to think about that, so just put it down to paranoia...[I] discovered [that]as well as raiding my flat they had raided my girlfriend's home and my family home. They could only have found out about my girlfriend and her address by following me. She has never been, although a vegan...known to the animal rights movement or the earth liberation movement, but [the police] still called around to her, asking her questions about me, so it was quite clear to me that they had followed me at the time, and also I had reason to believe that it wasn't Hants CID....They told my duty solicitor that I was being investigated by the Metropolitan Police...they raided the three addresses that I spent the most time at.
>
> (Molland in interview)

Molland was imprisoned in 1997 for his part in publishing details of direct action in *GA*, along with two other activists.

RTS have, typically, been subjected to an intensive state intelligence-gathering operation:

> RTS's office has been raided, telephones have been bugged and activists...have been followed, harassed and threatened with heavy conspiracy charges....A secret RTS action in December 1996 (an attempt to seize a BP tanker on the M25) was foiled by the unexpected presence of two hundred police at the activists' meeting point. How such information is obtained by the police is uncertain and can easily lead to paranoia...fear of infiltration, anxiety and suspicion which can themselves be debilitating.
>
> (*Do or Die!* 1997 (6): 3)

While repressive measures have increased as the anti-roads movement mobilised, over 1991–2 few arrests were made and support for direct action grew rapidly. 'Although hundreds of people have been arrested in the past 10 months in Britain's rapidly radicalising and growing environment movement, not one person has yet been sentenced' (*AU* 1992 (3): 7). Activist comparisons with Europe suggest that EF! (UK)'s key organisers perceived both repression and facilitation to have been operating at lower levels in the UK than on the Continent. Seven activists had participated in green movement activity in both the UK and Continental Europe, with many of the remainder showing a keen interest in the European green movement, confirming suggestions of the international scope of environmental direct action in the late 1980s and early 1990s (Welsh and McLeish 1996). Interview accounts suggest that the initial 'cost' of participation in direct action was perceived to be low, while the 'benefits' in terms of direct disruption, media attention, self-empowerment and solidarity creation were correspondingly high. In contrast, the authorities of Continental European countries were seen as reacting more repressively than the UK's to environmental direct action. Begg typically stated:

> I would not like to have to do all this direct action in, say, France with all the riot police there. In many ways Britain is a very tolerant country when it comes to dissent, and I value that tradition of tolerance. You know, since the Criminal Justice Act...[it has been] less so, but nonetheless it could be worse, let's face it.

Such an admission came despite the 'costs' that Begg had endured as an activist: he had narrowly escaped death when he was crushed by earth-moving equipment at Twyford, and was violently arrested on an EF! (UK) action in Yorkshire. 'Clare', who had participated in environmental direct action in the UK, Germany and Poland, further noted the existence in these countries of

> tougher police forces. It is pretty fucking easy to get arrested by the police in Berlin and you know the difference....It makes you not afraid of British policemen. Over there [the police] are seen as part of the state...searching

> everybody. [It] doesn't matter how liberal you look, and when you are nicked, you are scum, that's it.

EF! (UK)'s *Do or Die!* (1995 (5): 47) also noted high levels of Continental facilitation in contrast to the UK experience:

> One of the most interesting things we discovered was the way many of [Western] Europe's 'youth' environmental movements are funded. Their main source of income seems to come from government. When a network begins to grow the state gives them ridiculous amounts of money to organise endless seminars and alternative conferences. The most active in the movement end up spending all their time organising these things (and arguing about the money) rather than actually confronting those that are destroying the earth. This level of co-option is non-existent (as yet) in the student green movement over here – never mind EF!

'Lorenzo', who had been born in Italy and participated in environmental direct action in a number of European states, concluded that the lack of facilitation and the absence of severe repression meant that EF! (UK) found it easier to raise fundamentalist demands:

> it is all very mixed....[In] Belgium they have done very small road camp maximum of twenty people, something like the CJB has been in Europe much longer, CS gas [is used against protesters]...[yet] in Belgium government painted cycle lanes everywhere, Danish government [is] more sussed on the surface....We are being much more a pain in the arse here....Things talked about here are not talked about in European countries....We are just tackling everything.

Such examples seem to suggest that both facilitation and repression were used by European states to influence the green movement. While European states appeared more open to environmental policy demands and though environmentalists there might receive state funding, direct action as such was repressed. Such integration limited the use of NVDA, and groups were more likely to be either integrated or marginalised into violent action. As 'Clare' observed: the 'German EF! [are] hardcore....They just go and burn it down....[They are] not organizations but a few individuals....We are very lucky...over ground can work and we have shown that.'

The responses of 'opponents', in the form of counter-movements, police (including intelligence agencies) and private security firms, shifted from apparent apathy early on to a more aggressive stance as the anti-roads movement grew. Accounts of police violence, intelligence-gathering and conflict with private security guards become increasingly common in interview accounts of events after 1993. In common with the EF! (US) experience, UK state opposition seems to have sharply increased with the movement's growth. Although relatively mild in comparison to continental Europe, the increasing repression experienced in the UK influenced repertoires of action in three ways.

First, interaction between EF! (UK), the wider anti-roads movement and its opponents led to tactical adaptation. Increasingly, particular repertoires of action would be made ineffective by police or private security manoeuvres, and so would need to be replaced by new tactics. Typically, the initial mass EF! (UK) actions that involved occupations of timber yards and docks were neutralised as the firms targeted by EF! (UK) or the police gathered intelligence. Effective intelligence-gathering, which may have involved no more than reading publicly available leaflets promoting actions, allowed the police to mobilise in large enough numbers to make direct action difficult, while organisers were identified and served with injunctions. Finally the publicity material – posters and leaflets used to mobilise activists – allowed firms to discover the dates of actions and so close down for a day until the protest was over.

In response, activists developed micro actions such as ethical shoplifting; they also turned to promoting continuous action by setting up long-term camps. They began to avoid publicity in the media through which they could be identified in order to make difficult the serving of injunctions on individuals. In response, the DoT employed private investigators to photograph and identify anti-road activists so that they could be served with injunctions (Brown 1995: 6). On anti-road campaigns the – initially highly effective – tactic of 'digger-diving' was gradually neutralised; similarly, as the police and security services learned to counter tree-sitting, activists took to tree-dwelling, which, in its turn, was followed by the use of nets between branches. By January 1997, tunnelling underneath road constructions had become an effective and favoured repertoire. It is interesting that criminologists have identified an 'innovative cycle', in which interactions between criminals and law-enforcement agencies lead to ever-more sophisticated tactical adaptations (McIntosh 1971: 117; South 1988: 67). Such a model clearly fits both EF! (UK) and the wider anti-roads movement.

The second effect of repression was to increase sympathy for militant repertoires. Among interviewees, support for sabotage and self-defence strongly increased during the 1990s. While militants like 'Mix', 'Mark' and 'Clare' had a long-standing commitment to such stances, moderates were increasingly reassessing their approach. The costs of environmental direct action rose sharply both for activists and for their opponents. With increased use of police and security resources, together with legal changes, the risk of arrest and the cost of imprisonment increased. Sympathy did not result, despite rioting at Newbury, in the abandoning of mass non-violence and its substitution by sabotage. Sabotage coexisted as a repertoire, but was not seen by activists as 'violent'. Violence against persons was limited by the movement's tradition of non-violence and by lack of access to violent repertoires.

Third, repression resulted in 'costs' for opponents as well as for EF! (UK). Repression increased the profile of the anti-roads movement, increased public sympathy and promoted internal solidarity and identity. The very expense of policing environmental direct action significantly increased the cost of road building and other forms of activity contested by EF! (UK). Thus, the financial burden of security costs added to the roads bill, and in this way evolved into a tactical tool for activists!

Repression prevented the anti-roads movement from forcing road construction to be abandoned at specific sites. Yet the effect of repression was to both increase the financial cost of road building and to reduce the legitimacy of the roads programme by generating negative media images. The threat of protest may have been a factor in the reduction of the construction programme. Perception of such success, along with the success of tactical adaptations, seem to have sustained the anti-road movement's use of direct action.

O'Hara argues that intelligence-gathering and media disinformation are structurally linked. Access to police and intelligence services may be vital for journalists and may contribute to negative accounts of social movements. Yet attempts to delegitimise the movement via media accounts of violence seem to have been less successful in generating negative imagery than in the case of the animal liberation movement. It is possible to see in the numerous media reports a shift from the portrayal of EF! (UK) as a violent terrorist group to the mildly sympathetic depiction of the M41 occupation and the A30 evictions (compare Cohen 1992 and Moyes 1997 in the *Independent*). By 1997 'Swampy fever' manifested itself in the media adoration of the tunnel digger Daniel Hooper who, against his own intentions, became a public name with a tabloid newspaper column and numerous TV appearances. In turn, attempts by the state to act against radical greens were increasingly ineffective by 1997. In contrast, US radical greens, including EF! (US) and MOVE, have suffered almost unrelenting and, on occasion, murderous state pressure.

Mobilisation targets

Capital projects, such as roads, which influence movement activity, are the product of factors largely 'external' to those movements, and so can be included under Tarrow's definition (1994) of the POS. In this sense, political decisions and political institutions may shape not only the reaction to particular forms of grievance but also the level and nature of the grievance. As Rüdig and Lowe state: 'There would be no environmental movement...without the existence of environmental problems....This rather trivial assumption appears to be easily forgotten in the literature.' They claim that highly visible and disruptive capital projects 'such as hydroelectric dams and new airports have been central to environmental protest movements' (1986: 279). While the green movement may seek not merely to reverse particular grievances but to create more fundamental changes in structures and human attitudes, without such tangible grievances or 'mobilisation targets' green political critiques may appear too diffuse and abstract to achieve movement mobilisation (Rüdig 1993: 3). Such 'targets' act as condensation symbols or points at which particular concerns can accumulate, reaching a threshold where protest may occur. From a critical realist perspective, such 'targets' may act as the manifestations of 'real' yet invisible socio-environmental dynamics. The global environmental issues of the late 1980s and early 1990s, such as the greenhouse effect and potentially thinning ozone layer, have been regarded as such an invisible process (Cylke 1993: 26). Rüdig (1993: 3) has suggested that

> these types of issues do not offer the environmental movement any targets to mobilise people against. There are no grass-roots movements against the greenhouse effect, and not surprisingly so: global warming is an all-pervasive phenomenon which does not manifest itself in an easily recognisable form, its causes do not present themselves as individual targets, convenient for mobilisation purposes. Rather, it is an abstract issue; the perceived threat to the individual is intangible and those to blame are not easily identified...[consequently] the mobilisation potential for environmental political action has diminished.

Yet, despite such an opinion, the roads issue in particular proved a most evocative mobilisation target for EF! (UK). The expansion of 'the great car economy' was perceived as a manifestation of one of the 'invisible' causes of the apparent greenhouse effect. As such it indicated the failure of political 'realist' strategies to influence the UK policy process and that there was an ethos of political closure to green demands. Because roads have locally disruptive as well as potentially global effects, the opportunity existed for linking radical greens with more cautious local group campaigners to create sustained mobilisation.

Dudley and Richardson (1996: 26) believe that increases in anti-roads protest, using NVDA tactics, can be both correlated with and partially explained by peaks of construction in the late 1960s, early 1970s and early 1990s. This conclusion is clearly in line with the earlier survey of road-building and road protests (see Chapter 2), and helps to explain the renewal of the 1990s anti-roads movement. The *Economist* (19 February 1994: 27) asked the question: 'What explains the growth of the anti-roads lobby?', noting in reply: 'It is partly a reaction to the increase in road-building. Between 1987–88 and 1992–93 spending on the national network has risen by 52 [per cent] in real terms.'

Road construction in general, together with the late 1980s and early 1990s expansion, can be theorised in a number of ways. First, greens have seen it as socially and environmentally dysfunctional, and technocratically driven.

Second, pluralistic approaches stress the influence of a road lobby that includes consumer groups, some trade unionists and corporate industrial bodies.

Third, an effectively instrumentalist Marxist approach has stressed the connections between capitalist interests, most crudely those of major construction companies and the Thatcher government (Hamer 1987). In the early 1960s the Transport Minister Ernest Marples was a major shareholder in a civil engineering firm, which was awarded work on the Hammersmith Flyover, causing a scandal over such a perceived conflict of interests. During the 1980s and 1990s Conservative Party Treasurer Lord MacAlpine had a stake in a leading road construction company. Capitalism may be served in diverse ways by road building. Expansion of road transport can also be seen as a means of reducing the threat of industrial militancy. For example, during the 1984–5 miners strike, coal transportation was shifted to lorries rather than relying on the strongly unionised railways.

Fourth, structuralist Marxist approaches, in contrast, note the importance of car-ownership for strategies of economic accumulation. Such accounts suggest that far from being plotted by conspiratorial meetings of former public schoolboys, capitalism demands particular structural forms of change. Roads in short may lead to increased economic growth and may be championed by an economic logic of expansion supported not just by the road builders but by a range of interest groups.

Fifth, specific changes in economic accumulation also help explain changes in the rate of road construction. Time and again, since 1945, governments have planned ambitious construction programmes to accelerate the rate of economic growth, only to find that economic slumps have sabotaged their plans. The Lawson 'boom' of 1987–8, when the UK's economy rapidly expanded, may help account for the planned late 1980s–early 1990s increase in road building.

Sixth, cars may act as a means of establishing cultural identity: Gartment (1994: 14) goes as far as to use the term 'autopium' to describe their seductive charm. Cars have been seen as a means of establishing freedom, and critics of communal public transport are often condemned as collectivist and inflexible.

All of these accounts suggest that increased road construction is influenced not just by formal policy-making at state or regional levels but by a range of cultural and economic influences. As the earlier historical survey indicates, accumulation, discourse and protest have shaped road building.

There is evidence that the planned 1990s expansion of roads and car-ownership was linked to a growth of out-of-town superstores on greenfield sites. The car facilitates access to these supermarkets, where a whole host of diverse commodities can be bought. Often in negotiation with local authorities, supermarket chains offer to fund bypasses. Road-haulage firms, retailers and factories all provide pressure for increased road construction. One particular trend of the mid-1980s was the 'just-in-time' system, which was closely linked to increase in containerisation (Sayer 1986). Instead of storing large quantities of a product in warehouses, 'just-in-time' aimed to reduce the need for expensive storage by building goods 'to order'. It also reduces employment costs by removing store workers and – of greatest value to a firm – allows profit to be realised at increased speed.

> Factories using the system have many small deliveries a day (instead of a single larger delivery) from their suppliers, many of who [*sic*] in turn would be adopting the system. This amounts to using the road itself instead of a warehouse!...Major stores are increasingly moving to huge out-of-town sites and using less site space for warehousing; they use bar code scans at check-outs to determine which items are selling and have them delivered constantly from manufacturers and their own warehouses on other sites. These out-of-town stores choose greenfield sites not far from major roads to which they add service roads. Or, if they cannot find a big enough out-of-town site near a major road, they will offer money to a local council for a 'bypass' which will then be used by their lorries. This is what has been happening, for example, in Yeovil with a proposed Sainsbury's superstore.
>
> (Anon. 1994: 9)

Thus, from a green perspective, such out-of-town development has been perceived as funding new road construction, generating additional traffic and promoting a culture of car-ownership and consumerism.

Finally, the EC focus on the project of enhanced economic growth via the creation of a Single European Market placed pressure on the environment. The late 1980s and early 1990s expansion in the UK roads programme cannot be explained in its totality without reference to this fact. Many 'bypass' projects, overtly intended to reduce traffic congestion in towns and cities, have been seen by critics as incremental stages in new trans-European routes. The EC drew up plans in the late 1980s for upgrading Europe's principal strategic routes so as to enhance the unity of a Single Market and aid regions, such as Ireland, considered ripe for economic expansion. Plans for the UK centred around the Channel Tunnel at Folkestone, and eastern ports and routes that help to increase access from mainland Europe to Ireland.

> Road ministers talk of the individual elements of these roads (e.g. the A27, A35, etc., on the Folkestone–Honiton route), each being'improved' independently but in fact being massively upgraded and augmented in conjunction to accommodate (and encourage) freight lorries.
>
> (Anon. 1994: 9)

The ALARM UK anti-roads network argued that

> the Thatcher Government set out, literally, to alter the shape of Britain...through the positive encouragement of out-of-town retail and recreational developments, along with business parks and the largest road-building programme the country had ever seen...based on what Thatcher called 'the great car economy'.
>
> (Stewart *et al.* 1995: 7)

Roads provided a useful condensation point for framing global warming and other diffuse global environmental issues that also appealed to more narrowly focused local campaigns groups. Clearly such an issue allowed campaigners to frame their concern in such a way as to bridge the gap between different sorts of protester and maximise campaigning strength.

Allies and resources

EF! (UK), in acting as a catalyst for anti-roads action, not only attracted the direct participation of self-conscious greens but found allies in the form of local campaign groups and youth networks. Both sets of allies may have become more sympathetic to environmental direct action in response to political opportunity changes initiated or accelerated by the Thatcher and, later, Major governments.

In contrast to the cultural and social change demanded by green politics, local environmental campaign groups have been seen as content with modestly ameliorative reforms administered within a framework of local planning controls. Such groups traditionally have been thought to have strong network links with local councillors and others influential in the local planning process. Even the public inquiry system, criticised as unreceptive to green opposition to motorways, nuclear-power stations and other capital projects of 'national' significance, has appeared to be 'fairer' in dealing with local planning issues (Sieghart 1979). Acting defensively to protect cherished local landscapes from 'development', direct action repertoires may be seen as both alien and unnecessary to such amenity or conservation bodies.

Yet such groups approached EF! (UK) for support in a number of locations in the early 1990s. At Golden Hill in Bristol, between May 1992 and 1993, I took part in direct action to prevent the construction of a Tesco supermarket on a greenfield site. When construction work was due to start, the local Golden Hill Residents Association (GHRA), led by former Conservative Party activist Ian Martin, networked with green activists in Bristol, including disillusioned Green Party members like myself. Eventually, I was asked by Martin to bring in EF! (UK) activists to train local campaigners in NVDA techniques. For two months the site was occupied by campers. Mid-Somerset EF! (UK) trained the local activists and construction was resisted. Eventually, however, protesters were cleared and the store was built. The ELF distributed hoax £5 free-food vouchers when a Tesco store was opened in Didcot, Oxfordshire (Mercer 1994: 120). EF! (UK) disrupted the opening of the Didcot store and the Golden Hill store (where I was arrested). EF! (UK) also disrupted existing Tesco stores in Manchester and used direct action to prevent the construction of a Sainsbury's store in Yeovil.

Golden Hill indicates that local amenity groups were becoming frustrated with changes in the planning process, particularly the ability of the Minister of the Environment to overrule local decisions. Centralisation of planning, which began in the early 1980s and accelerated thereafter, seems to have made single-issue campaigners more open to direct action repertoires (Lowe and Flynn 1989: 262; Pye-Smith and Rose 1984). Such closure can be seen as one of the many effects of the increasing centralisation during three Thatcher governments.

GHRA's strong contacts with both the Conservative and the Labour Party in Bristol ensured that Bristol City Council refused planning-permission for the supermarket. Yet given the shift in planning power away from local government bodies, the developers were able to appeal to the Secretary of State for the Environment, who gave them permission to build. This appeal process was seen as unfair by the GHRA which had hoped that new Secretary of State Chris Patten would reverse the decision. When he did not, having exhausted legal policy channels, they resorted to a year-long campaign of NVDA. Without changes that centralised decision-making in the planning system it is doubtful whether the residents would have either felt the need to use NVDA or seen it as legitimate.

Although political scientists, sociologists and the media have often remarked on the links between 'conservative and middle-class' anti-roads campaigners and direct action protesters, this linkage is less surprising. Critics have long argued that the public inquiry system is biased against anti-roads and anti-nuclear power campaigners, noting the close links of inquiry inspectors to the state, the lack of funding for objectors and the limited nature of the questions that could be investigated (Sieghart 1979: 5). As already noted, this loss of legitimacy led the Conservation Society, the most conservative of the new environmental groups founded in the 1960s–1970s, and local anti-road campaigners to disrupt public inquiries in the 1970s. Some influential figures within the TDA and other local anti-roads groups have remained hostile to both EF! (UK) and direct action (Bryant 1996). Yet both Twyford and Golden Hill illustrate how a wider constituency from a more conservative (and often Conservative Party) background participated in direct action. Typically Hook (1993: 10) observed:

> It was interesting to note that many TDA members took part in the EF! protest, either bodily or by bringing food. When asked where his children were, one local, dressed in a Barbour and green wellington boots, replied that they were off 'tampering with some machinery; not to worry'.

At both Golden Hill and Twyford, EF! (UK) was invited by local campaigners to act as a freelance NVDA consultant, helping activists on the ground to sustain disruption. Tyme, using a different repertoire, had been brought in as a protest consultant by activists in the 1970s. Yet while the support of single-issue anti-roads campaigners for direct action was an acceleration of an earlier trend, the case of Golden Hill illustrates how even local conservationists with no previous direct action involvement felt that policy opportunities had closed and that direct action had become appropriate.

Youth culture

The 'rave' and 'acid house' youth subcultures of the late 1980s, revolving around dance music, open-air parties and the drug ecstasy (Collin and Godfrey 1997; Redhead 1993), were seen as increasingly apolitical and escapist both by participants and by academics (Jordan 1994: 113; Thornton 1994: 176). Yet both became targets of an ongoing 'moral panic', with politicians attacking drug-use and the promotion of disruptive parties (Collin and Godfrey 1997: 5). Legislation, culminating in the Criminal Justice Act (CJA) in 1994, criminalised many aspects of rave culture (Collin and Godfrey 1997: 227). Jordan (1994) notes how 'rave' as a movement was evolving from a situation of having 'no self-defined sense of itself as a collective in struggle', merely developing more elaborate means of evading the police to set up 'raves', to a more politicised form. In fact, by 1994 'ravers' established loose protest groups, similar in structure to EF! (UK), such as the Advance Party and the Freedom Network (Collin and Godfrey 1997: 227). These networks have enjoyed close links with EF! (UK), both nationally and locally, linking anti-road activism, cultural pursuits and the protest against the CJA. The Bill also threatened EF! (UK) and other activists, particularly Hunt-Saboteurs. Shane Collins, for instance, who created Brixton EF!, was a key network figure, providing the Freedom Network with offices at the squatted Cooltan's Arts Centre (Travis 1994: 5).

The CJA politicised many who may not have been previously involved in protest, and linked rave to environmental direct action in a project of common resistance to state criminalisation of such activities. Not all political participation took an unconventional form: in 1996, anticipating the General Election due to be held the following spring, Ministry of Sound, a large London dance club, sponsored a series of controversial posters promoting voter registration among young 'clubbers' (Armstrong 1996). EF! (UK) militants have argued that this link with youth culture gave rise to potential problems: 'the whole Criminal Justice thing...was very formative...in terms of an attack, [but] it brought in very naive people with no real previous involvement in politics, like the ravers and the Advance Party' (Mix in interview).

Activists such as Geffen and Freeman from the second cohort of EF! (UK) interviewees noted a strong crossover between EF! and rave culture. Rave music blasted through sound systems was a regular feature of Reclaim the Streets events in which I took part. While the green movement has often enjoyed youth-culture links (see Chapters 2 and 7), political factors accelerated the construction of links between rave and EF! (UK). In particular, what became known as DIY culture spanned the territory between these movements (see Chapter 7).

Freeman, in turn, noted how youth unemployment had resourced the DIY movement linking EF! and rave culture, arguing that DIY

> came about...mainly because there is so much unemployment and because you have so many people who are young and healthy and energetic, many of them well-educated. They have got skills, they have got enthusiasms and they have got no jobs, and they want something to do and they want to do something useful and important. Also, they can't afford to go to the films and the theatre and all the laid-on entertainment that people with money can do, [so] they put on their own entertainment, they have their own festivals, they have their own parties. They have their own actions and they do things like LETS schemes where you can trade without money. It could be all kinds of things. It could be art, it could be music, it could be protest, but it is doing-it-yourself without the kind of traditional money economy, and it is good fun.
>
> (Freeman in interview)

Such accounts challenge the post-materialist thesis, which argues that environmental concern is a product of rising prosperity, by seeing green activism as a means of surviving unemployment and poverty for the young. This point was also made by Victor Anderson who, as noted in Chapter 2, in December 1971 undertook direct action in Oxford Street, London, to protest against the effects of the car on the environment. Discussing why his group, Commitment, failed to create a mass environmental direct action movement, he observed:

> There was less awareness of the issues. Also I think unemployment has made a big difference, now there are a lot more people about with less to lose. At that time, when people were arrested in Oxford Street their main fear about being arrested wasn't that they would get fined...people were afraid of losing their jobs, whereas now if you have masses of people who don't have jobs anyway they will be prepared to go further.
>
> (Anderson in interview)

Thus, while some subcultural activities were made illegal, and so politicised, with elements becoming consciously counter-cultural, lack of employment opportunities lowered the relative cost of resisting the CJB using unconventional political action. In turn the anti-roads movement and rave culture were able to resource each other's protest activities and to develop their respective repertoires of action (Collin and Godfrey 1997: 227).

Conclusion

An enduring feature of the UK political system has been 'the relative closure of the British electoral process' to green and other left-libertarian parties. Unlike in many Continental European countries and Ireland, the Green Party had failed to secure seats in the UK's parliament by the 1990s, a factor that has tended to encourage non-electoral strategies (Rootes 1995: 79). To a limited extent, historical social-movement links to the Labour movement and the facilitation of environmental lobby groups may have reduced the use of direct action within the green movement.

Within this context, political factors in the late 1980s and early 1990s provided opportunities for the growth of an environmental movement which used repertoires of direct action and promoted grassroots internal organisation. The apparent opening of the POS to environmental demands, followed by its closure, between 1988 and 1992, are of primary importance. When such policy change stalled, the case for a network advocating participatory organisation and repertoires of direct action was strengthened among green activists. For the initial EF! (UK) activists this POS factor was mediated through existing environmental pressure groups and the Green Party. These groups became less confrontational, less participatory and, simultaneously, less successful in promoting green political goals. The anti-libertarian stance of these groups was based on the assumption that less-participatory structures would aid attempts to lobby for environmental policy change by discouraging confrontational and fundamentalist approaches to the state.

Equally EF! (US)'s co-founder Foreman explained his turn towards direct action in terms of the over-cautious strategies of existing environmental groups (Foreman 1991; Rucht 1995). A political factor contributing to the growth of EF! (US) was the election of an environment hostile politician in the form of Reagan as US President, and his appointment of the equally hostile James Watt as Secretary of the Interior (Scarce 1990: 21).

Political closure has been used to account for the growth of green movements in supposedly corporately governed states such as West Germany in the 1970s (Burns and van der Will 1989: 11; Scott 1990: 146; Steinmetz 1994: 195–6). The strongly neo-liberal state of contemporary Britain equally promotes such a phenomenon. It has been argued that Thatcherite assumptions were so invasive during the 1980s that realism within environmental pressure groups and the Green Party was accelerated by neo-liberal assumptions which shifted action from protest to forms of market choice.

> In many ways, the role of individual members in green parties and national environmental groups is similar to that of the green consumer. The only action required is that of the parting with money: you buy 'environmental action', in just the same way as you buy 'green' washing-up liquid. Naturally, the relationship between leaders and members becomes more similar to that between salesman and customer; personal participation in political action is replaced by paying money as the main form of interaction between green group and individual member.
>
> (Rüdig 1993: 5)

The collapse of the peace movement, the marginalisation of the left by 1990 and the growing 'political realism' in the Labour movement attracted new activists to the anti-roads movement once the initial mobilisation had occurred. Perceived low levels of facilitation and repression aided activism.

While Kriesi *et al.* (1995: 145) argue that movement activity is 'politically constructed', and as such is largely a product of both the shifts and the more-enduring POS features, changes in the scale and nature of specific grievances may also influence movement mobilisation. Grievances, too, increased with protest fuelled by an expanded roads programme, superstore development and other large-scale construction. Road construction provided a very material example of a cause of global warming and other forms of pollution that might be relatively invisible.

Additional economic and social factors aided EF! (UK)'s emergence and the renewal of anti-road protests in the 1990s. Changes in economic accumulation, accelerated by Conservative governments, increased youth unemployment. Unemployment and an expansion in student numbers created a pool of individuals with greater personal availability to participate in direct action. Via such economic effects and the continuing 'authoritarian popularism', an increasingly large and persecuted cultural movement in the form of new age travellers and other economic marginals, often with green links, has emerged (Lowe and Shaw 1993: 112–24). Equally, the contradictions in Mrs Thatcher's cultural projections that emphasised both England's traditional values, including conceptions of the 'green and pleasant land', and her neo-liberalism, may have contributed to the propelling of Conservative single-issue campaigners to greater toleration of greens (Rootes 1995: 81).

Political closure also accelerated the growth of local campaign group and youth culture support for environmental direct action. Discussing the participation of both 'radical green activists and local, often middle-class, residents', it has been claimed that anti-road campaigns convey the impression 'of a growing number of people who considered themselves to be disenfranchised from the policy process, and were prepared to go to almost any lengths to make their voices heard' (Dudley and Richardson 1996: 28). Changing political opportunities cannot wholly explain the arrival of EF! and the growth of anti-road campaigns in the UK, but do they point to circumstances that may have made direct action more attractive.

7 Culture, ideology and the anti-roads movement

> Hayduke thought. Finally the idea arrived. He said, 'My job is to save the fucking wilderness. I don't know anything else worth saving. That's simple, right?'
> 'Simpleminded,' she said.
> 'Good enough for me.'
>
> (Abbey 1991: 200)

> Heather sang along to the JAMM's *Jerusalem*: 'Bring me my harness and my rope/ Bring me my rock of kryptonite/ Bring me some flapjack and some dope/ Bring me a tarp for rainy nights/ I shall not cease from locking on/ Nor will my drum stand in my hand/ Till we have saved this place from them/ This England, green and pleasant land.'
>
> (Merrick 1996: 114)

Introduction

Green political theorists and activists have been fascinated by EF! (US)'s cultural practices and ideological assumptions. In the UK certain academics and sections of the media have shown an almost obsessive interest in the anti-road movement's use of music, poetry and art. Links with dance culture and connections with middle-class conservationists inspired by visions of a green and pleasant land have also been subject to much commentary. Culture has been used to exploit media interest and activists have set up their own forms of DIY communication using video. Reclaim the Streets actions are often giant parties with sound-systems, picnics and children's sandpits. Arriving at a road-camp, one can find the colours and unfamiliar sights disorientating.

Politics, in the form of ideology that criticises an existing society and promotes an alternative vision, especially perhaps green politics, cannot easily be separated from culture. A strong culture can bind activists together, creating a sense of solidarity which helps them to take part in 'high-cost' direct action, such as tree-sitting or tunnelling, that carries risk of arrest or even injury. Provocative 'condensation' symbols such as a road that threatens ancient down land, dramatised perhaps by story, myth or (as at Solsbury Hill) song, transform environmental issues from worrying but vague abstractions to realities that can be understood and engaged. Vibrant culture including music serves to make a movement visible and distinctive. Cultural forms, such as Abbey's novel of ecotage or camp-fire protest music, may make activism more enjoyable, cement activist identity, bring in new supporters or even transmit dramatic repertoires of action from one generation to the next. Finally, cultural transformation, particularly lifestyle-change, is a goal of much green action as well as an aid to movement mobilisation and maintenance.

Some academics have used the frame realignment model to show how culture can promote movement growth. Frame realignment is seen as a process whereby political activists so present a movement's beliefs that they resonate with a wide selection of potential supporters (Snow and Benford 1988a). Movements act as signifying agents, working to structure and shape meaning so as to motivate action-creating ideology. A 'frame' is defined as

> an interpretive schemata that simplifies and condenses the 'world out there' by selectively punctuating and encoding objects, situations, events, experiences, and sequences of actions within one's present or past environment.
>
> (Snow and Benford 1998a: 137)

It has been suggested that

> there are three core framing tasks: (1) a diagnosis of some event or aspect of social life as problematic and in need of alteration; (2) a proposed solution to the diagnosed problem that specifies what needs to be done; and (3) a call to arms or rationale for engaging in ameliorative or corrective action.
>
> (Snow and Benford 1988a: 199)

In contrast to the mass-media appeal of EF! (UK), EF! (US), it has been claimed, has failed to attract a large public following due to its inability to 'form adequate identification through effective appeals' (Jimmie and Palmer 1992: 20). In the US Swampy might have been 'framed' as a terrorist rather than an heroic tunneller. In the US Judi Bari, the EF!er who opposed tree-spiking, and MOVE were victims of bomb attacks, both portrayed by the press as perpetrators of their own assault.

'Frame realignment' can be criticised in much the same way as can the resource mobilisation and social constructivist approaches from which it is derived. Movements, more so than pressure groups, are motivated by political belief, they do not construct attractive forms of ideology simply as a means of resource mobilisation. It is easy to slip from a social constructivist perspective, examining how movement literature is written or video appeals are made, to the assumption that activists simply spin words and images in search of donations to fund alternative forms of career politics. Anti-road activists oppose roads, and while they may frame their opposition so as to increase active support, their goal is to combat the spread of concrete through greenfield areas. Radical greens seek to transform popular culture so as to create a greener society. Realigning, that is changing, their own views to reflect popular sentiments that are hostile to the construction of such a fundamental green hegemony, as a means of movement-mobilisation, would seem to undercut the very rationale and long-term political objectives of such mobilisation. EF! (UK) activists seek to transform belief not to court short-term popularity. In short, while frame realignment may be used to map ideology, it should not be used as a method which presupposes that ideologies from Leninism onwards are no more than the raw material of mail-shot campaigns.

Core framing

Activist accounts and texts do, however, indicate that EF! (UK) has carried out all three of the framing processes outlined by Snow and Benford – diagnosing the problem, advancing the prognosis and, finally, calling others to action via motivational frames. The results of all three processes were contested by new activists and the frames advanced shifted significantly.

All activists interviewed perceived there to be an ecological crisis of fundamental importance and global impact, which they felt was manifest in symptoms such as loss of biodiversity, greenhouse warming, ozone depletion and other ills. EF! (UK) co-founders Burbridge and Torrance emphasised deep ecology as a diagnostic frame, centring on the intrinsic worth of other species rather than looking exclusively at the human consequences of environmental damage (Burbridge 1992a and 1992b; Torrance 1991). Thus the first issue of *AU* stated: 'Earth First! is an international movement dedicated to preserving natural areas and fighting the forces of industrialism.' The press release publicising the first EF! (UK) action at Dungeness stressed the importance of a deep-ecology perspective, expressing concern about loss of plant diversity rather than the danger to human life of radiation leaks. Burbridge described a 1992 London meeting at which Murray Bookchin, the prominent US social ecologist and bitter opponent of EF!, spoke. Noting his disdain for such a diagnostic frame, Burbridge described the scene thus:

> In a spiritual fit of apocalyptic rage Murray launched into a vitriolic attack upon those whose approach to the impending eco-disaster he had an obvious contempt for. 'Thinking like a mountain??... Snails of equal worth to human beings?... Blah, I don't want to be eaten by an earthworm!!' he snorted. 'What juvenility!!' he screamed as he continued to liken all deep ecologists to nazis.
>
> (Burbridge 1992b: 8)

Despite this initial focus, Torrance believes that EF! (UK)'s diagnosis has shifted to take in 'social' issues, and he contrasted this shift with his perception of EF! (UK):

> EF! in the States...were coming from a biocentric equality for all life, deep-ecology, point of view...rather than the deep social-change issues. I still don't think Earth First! in the United States is in anyway as social as it is over here....In the States I think it's more coming from deeper ecological roots, seeing yourself as part of the world....It really started as a wilderness preservation movement, yeah, a radical movement to protect the last areas of wilderness and to reclaim some areas of wilderness.
>
> (Torrance in interview)

Several interviewees were hostile to what they perceived to be an EF! (US) diagnosis, one which they felt rejected the social dimension. Typically, in interview, 'Clare' noted:

> Dave Foreman is a well-dodgy bloke. You hear some really strange views, like some of them are really conservative, right-wing in a way...They are there for the wilderness...[but] you have to be there for people [also]. They are just as much a part of the earth....I think it is a very different movement.

Lush argued that 'EF!ers are very aware of a lot of social issues about equality, about [the] taking into consideration of minorities, of listening to people, of sex and gender issues'. Such considerations were derived by activists both from the UK green movement and from US political repertoires. Feminists who brought with them the experience of Greenham sought to introduce social goals to EF! (UK)'s diagnostic frame, as did militants influenced by anarchism. Judi Bari, from the EF! (US) 'social-justice faction', the MOVE organisation and even Bookchin were cited by activists as influential North-American sources of diagnostic framing. Typically, Garland was supportive of both Bookchin, who he met and praised in May 1992, and Bari, whose essay 'The Feminisation of Earth First!' he strongly promoted.

Different prognostic frames were voiced by EF! (UK) and other anti-road activists, variously attacking industry, the state, reductionist philosophies and over-population as causes of ecological and social ills. Deep ecology, particularly when associated with misanthropic statements stressing a neo-Malthusian concern with population increase, soon ceased to be important. Malthusian comments were relatively rare in activist interviews; and in this context interview bias should be considered, as I had often stated in print (see, for example, Wall 1990) my rejection of 'over-population' explanations and some interviewees were aware of my views. Nonetheless, Noble argued that 'the population crisis is so bad that we may have to stop having children, or only have one child', while similar sentiments were articulated by Tilly and 'Mary'. One prominent activist I met claimed that he intended to have a vasectomy and rather alarmingly claimed to be urging other young campaigners to combat their fertility in the same way. Yet discussion of population is absent from *Action Update*, is rarely mentioned in other EF! (UK) publications and is missing from the anti-roads literature. The contrast with EF! (US) is dramatic: into the 1990s and the US *Earth First! Journal* advertises bumper stickers opposing pregnancy.

A number of alternative prognoses to the population bomb were advocated by activists in interview. Some were strongly critical of technology and sought to restore the earth to a pre-scientific or even pre-agricultural state. Mix observed:

> I think a return to the Palaeolithic [Age] would do it. As I say, I think mass-organised society has to be destroyed and we have to decentralise down into small-scale, self-governing, self-sufficient communities, free of, like, hierarchy and alienation, as existed in the Palaeolithic.

Serious academic accounts of the flawed genesis and social costs of both agricultural and technological development suggest such views cannot be dismissed as entirely naive (Mumford 1970). Another activist argued, with rather more historical restraint:

> Only [with] nineteenth- [and] twentieth-century technology have we been able to do these things....Some people say it is money or greed, but we have always had greed...we have always had all these problems with ourselves, but *we* have more power. I don't know what we do about disinventing all this stuff....I suppose I am a Luddite. When you look at any new invention you have to think it has at least twice as many bad things wrong with it as good things....I personally would not have electricity....I have been to villages in Africa with no electricity, and they are fine, they are great. Happiness does not depend on stuff.
>
> (Freeman in interview)

Such 'regressive' views were not held by all the activists interviewed: while criticising consumerist excess and technological 'overdevelopment', Hunt, for example, argued:

> I would like EF! not to be associated with primitivism....There is a danger that environmentalism can be quite a conservative movement. I think it is quite important that EF! aligns itself with, sort of, progressive social movements in terms of feminism and anti-racism, so I would like to see the forward-to-Earth or back-to-Earth combined with progressive social poli-

> cies. . . . I would like to see more access to collective resources, like laundries, that the whole community could have access to. . . . Everybody thinks we should own everything individually, so there is a lot of scope for environmental salvation.
>
> (Hunt in interview)

Industrialisation, capitalism and the state were common, and often conflated, diagnostic targets. Those most sympathetic to deep ecology stress the dangers of 'industrialisation' and an anti-biocentric or reductionist philosophy of life. Such an approach is a common diagnostic frame within the UK green movement, but in EF! (UK) is linked particularly to the left. Durham saw his vision as 'eco-socialist and anarchist'. In turn, a leaflet distributed by RTS during the 1996 M41 street party noted: 'the streets are as full of capitalism as of cars and the pollution of capitalism is much more insidious'. An article in *Do or Die!* (1997 (6): 2) argued that

> cars are just one piece of the jigsaw and RTS is about raising the wider questions behind the transport issue – about the political and economic forces which drive 'car culture'. Governments claim that 'roads are good for the economy'. More goods travelling on longer journeys, more petrol being burnt, more customers at out-of-town supermarkets – it is all about increasing 'consumption', because that is an indicator of 'economic growth'. The greedy, short-term exploitation of dwindling resources regardless of the immediate or long-term costs. Therefore RTS's attack on cars cannot be detached from a wider attack on capitalism itself.

Hostility to the state and to capital are most obviously conceptualised as ideological repertoires drawn from green movement anarchists like Bookchin and networks such as *GA* and GL. Again Judi Bari was influential in this context.

In contrast, local anti-road campaigners from rural areas, sometimes, like Twyford's Barbara Bryant, with a background in Conservative Party politics, are less likely to articulate anti-capitalist and anti-state views. Bryant (1996) argued that privatisation could have funded a toll-road beneath the Down. Many environmental pressure groups, including Greenpeace, have sought co-operative partnership with industry and even government agencies. *Do or Die!* has condemned such links and challenged what it sees as corporate environmentalism:

> Of course, such 'greenwashing' is rife. . . . Greenpeace's involvement in the development of fuel-efficient cars, the Shell 'Better Britain' campaign, the Ford Conservation Awards, Esso Tree Week, Tarmac's astonishing, sanctimonious conversion to 'environmentalism' (pass the sickbag) – the list is endless.
>
> (1996 (6): 22)

Motivational frames also brought in a strong social element, with interviewees arguing that ecological problems threatened human survival, and 'frame-bridging' to link ecology to wider goals of human emancipation. Laughton, who attempted to create an EF! (UK) movement in 1987, argued that a poor motivational frame, linked to EF! (US)'s apparently conservative deep ecology, made his task difficult:

> A lot of EF! analysis is anti-human...[and a] very important part of [that] analysis is that human beings are messing up the planet and should keep off it....I realised that taken to EF!'s extent [analysis] also gets you into very dodgy territory because you have to question what the hell you are doing it all for....I am doing it for me, and I am a human being, and doing it for friends who are also human beings. Yet we are also talking ourselves out of existence. I could intellectually agree, but I couldn't really get motivated.
>
> (Laughton in interview)

Thus, in contrast, a typical framing statement from EF! (UK), taken from *Wild*, linked diagnosis, prognosis and motivational frame, addressing those

> who are no longer content to sit around while the corporate greedheads and their puppet politicians destroy the world. More and more people are becoming empowered by confronting the rot face on and winning! It is only when people confront today's problems with direct action that the situation will change, but it has to happen now, or else say goodbye to your future and our home, The Earth.
>
> (Anon. 1992: 2)

Such a statement links the core framing tasks to the notion of empowerment and lays the stress on the need to undertake direct action. Examination of EF! (UK)'s approach to core framing suggests a shift from deep ecology to more inclusive forms of framing that emphasise social issues. Equally, the wider anti-roads movement, as we have seen, stresses the social as well as the environmental ill-effects of the car. The deep-ecology frame, if taken to mean an exclusive focus on non-human nature and the incorporation of conservative neo-Malthusian political statements, had ceased to be important within a few months of EF! (UK)'s foundation in April 1991 – for three reasons.

First, it 'played' badly among the green activists initially recruited by EF! (UK). Green repertoires that focused on ecological crises were well established in the UK and embedded in wider society by the early 1990s. Those who, from the early 1980s onwards, self-consciously promoted a green movement specifically sought to link environmental issues, decentralisation, non-violence and social concerns (Wall 1994b).

Second, the perceived lack of UK 'wilderness' made it difficult to use elements of deep ecology in a motivational frame. While it is possible to criticise EF! (US)'s wilderness assumptions on both ecological and social grounds, the concept of wilderness seems easier to sustain in a North American context (Daniels 1994; Soule and Lease 1995). Indeed the 1964 US Wilderness Act defines 'wilderness'

> in contrast to those areas where man and his works dominate the landscape...as an area where the earth and its community of life are untrammelled by man, where man himself is a visitor who does not remain.
>
> (Borgmann 1995)

The UK's landscape is more obviously socially constructed, although, of course, as a critical-realist approach might suggest, under relatively enduring natural conditions and as a result of equally concrete social forces. The social myth of the British countryside is based on a landscape created largely during the last 200 years by processes of the enclosure of common land. Individuals dispossessed by enclosure provided the workforce for industrial production in rapidly growing cities. The landscapes of large country estates, often praised by the Romantic poets for their natural beauty, were also constructed by class forces. As Short (1991: 66) notes: 'The English countryside...is a landscape of power whose "mythic" properties are comparatively recent in origin.' Since the Neolithic Period human societies in the islands of Britain and Ireland have been constantly revolutionising non-human nature (Evans 1975: 108). Even without human intervention non-human landscapes would not have remained static, with Ice Ages, for example, transforming the landscape regularly, the last as recently as 10,000 years ago (Evans 1975: 2). While equally dynamic social and natural forces have acted upon the landscapes of North America, the sub-continent's apparent vastness and alien appearance to commentators such as Muir and Leopold has made notions of wilderness easier to sustain. As Torrance, noted: 'I think...the very culture and nature of Britain are very...different from the [United] States'. There's no wilderness and the whole culture, say no more, is vastly different.'

Third, even in North America activists and theorists have increasingly found it difficult to understand the origins of perceived environmental problems without reference to social notions, including culture, economic change, social class and state power. MOVE, for example, sees a unity between human and non-human nature and believes that social injustice and ecological problems have similar causes (Friends of MOVE 1996: 3). The beliefs of both Bari and MOVE suggest that a distinction can be made between social and asocial forms of deep ecology. While both reject anthropocentric approaches, asocial deep ecology ignores the fact that ecological destruction is the product of particular social practices and institutions, or at least is mediated through them. Equally, while neither Bari's approach nor that of MOVE rejects the intrinsic worth of other species, both do also embrace human-justice issues (Bari 1992; Friends of MOVE 1996).

In this context, Dobson argues that a contrast may be made between 'weak' and 'strong' forms of deep ecology. While the latter rejects 'human instrumentalism' by valuing non-human nature, unlike 'strong' deep ecology it does not remove the human species from its concerns (Dobson 1990: 63–5). Yet without a particular stress on the essential value of wilderness and the deprioritising of human goals it can be difficult to distinguish deep ecology from wider green political ideology. Indeed commentators may argue that 'deep ecology informs green politics' in general (Bennie *et al.* 1995; Dobson 1990: 47). As defined initially by the Norwegian philosopher Naess deep ecology's formulation appears similar to the social ecology with which it is often contrasted: 'Ecologically responsible policies are concerned only in part with pollution and resource depletion. There are deeper concerns which touch upon principles of diversity, complexity, autonomy, decentralisation, symbiosis, egalitarianism, and classlessness' (1973: 95). Rather than situating either EF! (UK) or the wider anti-roads movement of the 1990s in terms of deep ecology, the issues of participation and direct action espoused by both seem far more important. From Green Anarchists to Conservative Party activists like Bryant, anti-road campaigners have been united in their resistance to political closure.

Participation

Burbridge (1991: 10) typically argued:

> Progress is finally being made as concerned individuals free themselves from the disempowering and patronising behaviour of certain groups, to establish a movement that works from the bottom up...where anger and love play as important a part as depressing statistics and expensive reports....This is a movement that is truly green, decentralised and self-empowering....Green bureaucrats, move over! The real green movement is on its way.

Consequently far less emphasis was placed on other elements of ideology as distinguishing features. Lush stated when interviewed:

> It doesn't have one big belief system....People congregate under the EF! banner rather than an FoE banner because they believe in NVDA, they are revolutionary rather than reformist, they are anarchic and don't believe in government.

This anarchic and open emphasis has remained largely stable since 1991. For example, issue 33/34 of *AU* (December 1996) noted that EF! (UK)

> is not so much a cohesive group or a campaign as a convenient banner for people who share similar philosophies to work under. The general principles behind the name are non-hierarchical organisation and direct action to confront, stop, and eventually reverse the forces that are destroying the planet and it's inhabitants.

An earlier *AU* (20 September 1995) description noted:

> Earth First! is a network of independent groups who believe in doing just that: putting the earth first. It is based on a concept of non-hierarchical organisation, direct action, and the empowerment of individuals to confront the ecological catastrophe facing our planet.

Freeman observed that EF! (UK) 'believes that individuals can take action...not just pay their money'. The majority of interviewees describe EF! (UK) as a 'network', suggesting that this term has a particular attractiveness. 'Mary' suggested that 'the only sensible words anybody has come up with to describe it [are]..."information network" and..."activist network" '. EF! (UK) was described variously by activists as 'decentralised', 'loose', 'chaotic', even as a form of 'disorganisation' . One founding activist went so far as to argue:

> Basically, it is not [organised]. EF! is about people doing actions, if you have got a concern and carry out an action then you are EF! There is no central command structure or even network. It is about people feeling strongly and doing the actions. So people see their local marshland is going to be dug, and they sort of feel strongly against this and they use civil disobedience and direct action means to stop this, and they have the right to call themselves EF!
>
> (Durham in interview)

Movement texts illustrate how EF! (UK) functions as a 'network of networks', providing a series of 'entry points' for would-be activists by virtue of the widely distributed 'contact list', a feature of every EF! (UK) publication since 1991. For example, issue 33/34 of *AU* (December 1996) states:

> The contacts are autonomous groups that consider themselves to share [certain]...principles. International contacts are entry points into the country's [*sic*] network of groups. Other contacts are ones whose aims and principles are [those] that we agree with fully or in part, or groups that we think may be of use or interest to people.

It is clear that EF! (UK) adheres to the series of six 'left-libertarian' organisational characteristics identified by Kitschelt (1989: 67) in his analysis of European green parties:

> 1 There are no barriers to membership; formal requirements such as membership cards, dues, endorsement of an ideology or a probation period are absent.
> 2 Citizenship rights in the movement are primarily exercised through 'presence' and participation at gatherings, not through representative institutions.
> 3 Organizational statutes are rudimentary or nonexistent. There are few formalized decision procedures.
> 4 There is little division of labor and activists are 'amateurs' who change tasks and roles frequently.
> 5 Elite positions are severely circumscribed in authority and tenure. Spokespeople and public representatives of the organization are rotated frequently.
> 6 Activists show little organizational loyalty or attention to the goal of organizational maintenance.

That EF! (UK) adheres to these characteristics is supported by both interview data and movement texts:

1 Formal membership does not exist; instead, supporters take part in direct action and many subscribe to EF! (UK) publications. An activist who helped produce *AU* noted: 'It is a network of autonomous groups...linked by an informal network of people who know each other. It has no head office, no official hierarchy, [and] there is no membership.' Initially, EF! (UK)'s organisational form was contested, with some activists arguing the need for formal structure, while the first national EF! (UK) Gathering defined a set of exclusive conditions for participation in the network (see pp. 57–8). Since 1992 Gatherings have made fewer formal decisions and a much looser approach to national organisation has prevailed.

2 Decision-making at national Gatherings and at local EF! (UK) meetings is informal or based on agreed decision-making structures.

> I didn't realise we were doing consensus decision-making until we had been doing it for a long time, but on the M11 we never voted. Rule number one is you never vote, you keep discussing until people agree to agree or agree to disagree. It wouldn't work where you have people who really are opposed to each other, [but in EF!] everybody is fairly in agreement to start with....There isn't a rule book. Occasionally we have

> to vote on things. We had this one guy who used to challenge the whole principle who was quite mad, but he was not so mad you could say [that] he was mad, and kick him out.
>
> (Freeman in interview)

3 There is no national EF! (UK) office; there are no national officers; and the only 'institutions' of EF! (UK) are *AU*, a number of less-formal publications and annual national Gatherings. Thus the first *AU* stated: 'We employ no one and have, so far, avoided the bureaucracy and elitism so prevalent in other green groups.'

4 While *AU* is produced by a strong and well-resourced local group, publication rotates on a regular basis, with the intentions of dissipating concentrations of power and instead of empowering individual activists. While the earliest *AU* editorial groups, Lea Valley (London) and Oxford, were criticised by militants for taking a leadership role, the editorial function has shifted to a new group on several occasions since and such criticism has become more muted. Discussing the M11 campaign, Freeman noted:

> Stewart, who is an excellent chair, chaired the M11 for a while, until he was called the 'leader of the M11' in an article, and from that time on he was never chair again. It was just so wrong....[Now] the chair is revolving.
>
> (Freeman in interview)

5 Leadership roles have been rejected, and local groups do not seem to elect formal officers as such. Nonetheless Lush, among others, noted that 'democracy is a weird word. There are strong charismatic people in every group, and there are always hidden hierarchies in non-hierarchical things.' One individual argued that Manchester EF! was dominated by a small number of student activists, and individuals outside of their circle felt excluded (Allen in interview). EF! (UK) activists, particularly those with a more 'militant' perspective, have been critical of the role of those media spokespersons who have emerged within the broader anti-roads and direct action movements.

6 Little effort has ever been made to recruit new activists or build an organisation. Even during 1991, when Burbridge and Torrance sought to create a movement, the emphasis was on the importance of local EF! (UK) activity and autonomy. Yet, it is clear that actions have, on occasions, been promoted as specific EF! (UK) national events, so as to cement the minimum movement identity necessary to prevent total biodegradation.

In contrast other UK green groups tend to have formal membership requirements, control over local groups, a national office, identifiable leaders or spokespersons, and they work hard to maintain organisational stability and membership loyalty. Even the green political organisations which refer to decentralisation as a key element of their ideology seem highly formal in comparison. For example, *GA* and GL articulate distinctive political programmes which they promote to would-be supporters.

EF! (UK)'s strong emphasis on organisational matters distinguishes it from much of the green movement, but allows it to make informal links with diverse groups and other loose networks. Concern with organisational questions provides a means both of boundary maintenance and of bridging weak ties to mobilise resources. Social movement theory suggests that adherence to a loose structure allows for swift growth during the emergence of a new campaign.

EF! (UK)'s organisational 'purity' can be seen as a reaction to green movement political realism and wider social-movement professionalisation. A perceived gap between green ideology and the organisational practices of the UK green movement led activists to EF! (UK) and encouraged much of the wider anti-roads movement to reject conventional structures. A wider 'democratic deficit' in the UK and frustration with conventional political activity have made EF! (UK)'s approach popular among young radicals from outside the green movement.

EF! (UK) activists indulge in frame-bridging to link their organisational outlook and views on direct action. Direct action is seen as a way of bypassing centralised power structures and the processes of representative democracy. It is thus identified as a way of gaining influence over the political process that is not mediated by parliament, pressure groups or the mass media. Thus, 'left-libertarian' organisational characteristics can be interpreted as a product of strategic choice (Kitschelt 1989: 66). Extreme forms of decentralisation are also seen as ways of avoiding police action. Hunt argued that EF! (UK) 'can take on certain underground activities because of its fluidity. [Because] it has no formal membership...it has no assets to seize, or offices. It is in quite a strong position to carry out work that the national organisations cannot do.' Such an organisational approach can be seen as prefiguring the norms of a green society and, as such, a means of promoting new 'codes', as Melucci (1996) has argued.

EF! (UK)'s organisational fluidity was deliberately chosen. Yet, rather than being constructed from 'scratch', it was built from existing green repertoires. The Green Student Network (GSN) clearly influenced EF! (UK), as did the experience of Greenham and ALF activists:

> Down at the Dongas...things were organised largely as a result of the Greenham women who were very good at sorting the situation so that everybody had a say, so that it was collective decision-making affinity groups.

> They basically...created a very strong grounding. It was anarchy action, and it worked.
>
> (Jazz in interview)

The GSN provided an organisational repertoire that included 'networking', consensus democracy, rotation and the use of gatherings, plus network newsletters to foster the minimal identity necessary for network survival. In turn, the radical animal-liberation networks indicated that autonomous local groups are valuable as agents of illegal activity. All three bodies (the GSN, Greenham women and the ALF) provided activists who brought organisational repertoires with them to EF! (UK).

Rucht (1995: 75) maintains that in respect of organisation EF! (US) closely resembles EF! (UK):

> It is difficult to get a clear picture of how Earth First! is structured and how it functions internally. This has less to do with the fact that radical supporters of Earth First! have also committed acts of sabotage and have been forced underground than that a pronounced sense of anti-institutionalism predominates. Earth First! just does not want to be an organisation.

In contrast, the wider anti-roads movement, while dominated by similarly loose networks such as Road Alert and ALARM UK, includes SMOs such as FoE and Transport 2000. In turn, local amenity groups, with chairs, secretaries and the like, may be structured conventionally and formally.

Direct action

Activists differ in the degree of emphasis they place on the role and significance of direct action. Direct action may be an element of a 'radical-flank' process, where perceived 'extremists' provide pressure that legitimises 'moderates' and encourages shifts in public perception and policy change (McAdam *et al.* 1996: 14). For example, it has been claimed that in the US 'the seemingly outrageous or illegal acts of groups like Earth First!...make the demands of the Sierra Club or Audubon Society appear reasonable' (Zisk 1992: 36; see also Taylor 1991: 262). Activists endorsing the radical-flank conception see direct action as largely symbolic, acting to legitimise the existing demands of environmental pressure groups and radicalising them by providing competitive pressure for media attention. In turn, it may act to delegitimise road construction and encourage greater activist involvement. In interview, Torrance noted:

> The first time Earth First! made national news with the little Earth First! flag in the corner of the screen was over Twyford...and it was...the start of the

> other groups really seeing that direct action could get you fucking publicity – it could get you heaps.

Explaining their approach, exponents of radical flank, such as Marshall, talk of a green movement 'spectrum', from the Earth Liberation Front at one 'extreme' to the environmental pressure groups at the other. EF! (UK), by carrying out NVDA and projecting an 'extreme' iconic presence, exerts radical-flank pressure. It gives environmental pressure groups a more reasonable appearance and thus helps them to achieve greater success in preserving the environment. Marshall noted when interviewed in 1995:

> I am now delighted with the way things are going...FoE are now doing NVDA training with all their local groups, [and] Greenpeace have now said they are going to do it. What I wanted to see was a situation where FoE in particular moved towards doing that kind of Gandhian, that very controlled, safe, form of NVDA, and that would then move; and the next stage on the spectrum would be, kind of, Earth First! 'hairies', and the next stage on the spectrum would be, say, ELF, and then it would all work. It would all work together.

Militants stress that direct action should be applied with disruptive intent. Such direct action bypasses mediation, such as articulating with other environmental pressure groups and the media in order to influence the state. Advocates of non-mediatory direct action have aimed to increase the economic cost of road building. The practice of militant disruption vitally politicises participants, encouraging them to act rather than to represent their demands to mediating institutions, such as the media or a pressure group. Advocates of this approach were sympathetic to sabotage and rejected an exclusive focus on mass NVDA. Typically, Mix argued that direct action should focus on

> a rejection or supercession of 'the system' rather than trying to negotiate or win accommodation with it. This, of course, is directly contrary to the line the green lobbyists and (in disguised form) the Gandhian NVDAers were pushing. In this context, the point about direct action is that it ends representation, delegation of power and division of labour, and in this it prefigures anarchy – damage is only an issue inasmuch as it is 'two fingers up' to the 'lobby-with-your-arse' brigade so big in the 1980s.

The rejection of representation has been portrayed as the 'central theme of anarchism' because 'political representation signifies the delegation of power from one group or individual to another, and with that delegation comes the risk of exploitation by the group or individual to whom power has been ceded' (May 1994: 47).

Non-mediatory and radical-flank approaches were strongly advocated by a minority of interviewees. Others were less reflective, or less certain about the political assumptions behind their actions. Alternative accounts of the importance of direct action were articulated, and non-mediatory and radical flank assumptions were often mixed. In this context Torrance noted:

> Economic pressure, I think, [is] a real card Earth First! has played all around the globe. By using some of the other things you have talked about, like physical disruption, media, whatever, it has exerted great economic pressure – so much so that some companies [which] have been logging some areas have had to pull out. Economic pressure in terms of monkeywrenching, monkeywrenching to fuck any of their equipment if it comes anywhere near, by initiating campaigns so their sales drop, etc.... And I think, really, at the end of the day EF! is about using the effective tools in the toolbox, whatever's there... short of violence.

Equally, demands for autonomous space were important to many interviewees, who saw direct action, especially when it involved the continuous occupation of a particular space, as a means of creating a zone free from state intervention. The construction of 'temporary autonomous zones' provides space for non-mediatory action (Bey 1985) and, paradoxically, powerful iconic images for media consumption. Promoting non-mediatory action, RTS distributed a leaflet at the 1996 M41 street party, stating:

> We are basically about taking back public space from the enclosed private arena. At its simplest it is an attack on cars as a principal agent of enclosure. It's about reclaiming the streets as public inclusive space from the private exclusive use of the car. But we believe in this as a broader principle, taking back those things that have been enclosed within capitalist circulation and returning them to collective use.

Nonetheless the different forms of action undertaken by individual activists within EF! (UK) may be underpinned by dissimilar and, occasionally, incompatible assumptions. As I have argued in Chapter 3, the bitter disputes between some EF! (UK) activists over contending repertoires of action were products largely of conflicting green-movement hybridisation. There were those activists, predominantly but not exclusively from the peace movement, who saw EF! (UK) as a vehicle for mass NVDA of a primarily symbolic kind; others, predominantly but not exclusively from animal-liberation networks, saw EF! (UK) as a focus for more disruptive and non-mediatory action. Do you silently creep up to the cherry-picker and remove engine parts? Or do you ritually cut a single strand of wire from the compound and give yourself up to the authorities? Both forms of action have brought results, but each rests upon its distinctive understanding of the politics of direct action. Action repertoires are not just the products of individual political philosophies: they are hybrids developed out of earlier polarities, with the influence of protest policing forcing competitive adaptations (see p. 130).

Frame-bridging and diffusion

It is clear that frame-bridging has occurred, allowing links to be made between anti-roads and animal liberation networks.

> Basically Earth First! is about putting the Earth first. Its philosophy is the Earth must come first, and even if a development that is bad for the planet gives short- or medium-term benefits for humans, in the long run it will be detrimental to humans because of the nature of ecological destruction.... It is also about recognising other species on the planet that have an equal right to life. It is not about 'What are they worth to humans?' They are alive, they exist, they have the right to exist, like women don't exist to be tools of and to give pleasure to men. Humans must recognise that we are one species amongst other species and we don't have a right to dominate others.
>
> (Molland in interview)

Networking and POS factors, such as the introduction of the Criminal Justice Act (CJA), introduced new issues in a process of multiple frame-bridging. By 1996 EF! (UK) activists had extended their diagnosis to include direct action in support of Labour movement struggles such as those of the London Underground train drivers and the Liverpool dockers (Pilger 1996b). Ultimately, despite such extensions and an early deep-ecology concern, there seems to be little to distinguish EF! (UK)'s diagnostic frame(s) from those of the wider UK green movement or the social-justice faction of EF! (US).

Contradictions between the different assumptions of anti-road allies were apparent, as campaigners at Newbury noted: 'Stranger still was the local fox hunters coming down on their horses to take direct action against the road that would cut through their hunting land, protesting alongside activists many of whom were hunt sabs' (*Do or Die!* 1997 (6): 25).

Despite the difference in orientation between EF! (US) and the UK's green networks, EF! (US) did provide a potent symbol, attracting both green-activist attention and media coverage. As critical realism suggests: 'Things that are believed become real' (Miles and Huberman 1994: 5), or at least have very real effects. EF! (US), while it proved a relatively minor influence on EF! (UK)'s political assumptions, enjoyed considerable symbolic currency with many green activists, the green media and the mass media in the late 1980s and early 1990s.

Such symbolic currency aided mobilisation. Begg noted: 'I got a leaflet...from Hastings Earth First!...I knew the Earth First! name from having heard about American Earth First!, and it interested me straight away.' Rather than acting as a source of distinctive political assumptions or novel practices, EF! was a known 'brand-name', providing a set of symbols around which activists could cohere and build a movement.

DIY culture

EF! (UK)'s trajectory brought it into contact with the DIY culture, a loose movement that linked anti-authoritarian politics, youth dance culture and environmental protest (see Chapter 6 and Collin and Godfrey 1997: 226). This contact led to the development of new tactics and the construction of an enlarged constituency of anti-road activists.

DIY has its origins in traditions of subcultural self-help, including the 'Blues' party and sound system tradition, 'positive punk' and 'new age traveller' lifestyles. 'Sound systems' have been described as 'simply loose affiliations of young people who organise and supervise parties and raves' that originated in Jamaica in the 1960s when the government-controlled radio stations refused to play 'the diet of black Blues ordinary people wanted to hear' (Anon. 1993). Sound systems were then brought to the UK by the continuing Jamaican diaspora of the 1960s and 1970s. The remarkable success, at least in cultural terms, of Rastafarianism and its affinities with green politics are subjects too important and large to be discussed here.

Punk, a subcultural youth movement of the 1970s, stressed its nihilistic opposition to authority and the commercial music industry. 'Positive punk' was

> very much aligned to the new social movements. It was anarchistic, non-compromising, [and contained] many aspects of EF! (UK). Really, it was. It made a big effort to encourage political awareness, responsibility and activism, and published lists of groups like hunt-saboteurs and animal liberation and cruise-watch and the peace camps. And there was also a big aspect of carrying on the tradition of do-it-yourself punk that the mainstream punk movement had lost somewhere along the way. It was non-elitist...[and] about participation, not being a spectator. So that is quite an attractive part of it.
>
> (Hunt in interview)

A third ingredient of DIY can be found in 'urban autonomism' (Anings 1980; Kramer 1988; Kriesi *et al.* 1995). This counter-culture bases its identity around its opposition to prevailing society, 'squats' unoccupied buildings as both a material act of combating homelessness and a symbolic challenge, violently opposes both the state and far-right groups and espouses anarchist sentiments. In

1985, linking 'autonomism', 'positive punk' and 'sound-system' influences, a group of new age travellers, the Mutoid Waste Company, 'so called due to their predilection for customising their vehicles into futuristic monsters and making junk sculpture at festivals', squatted a disused bus depot in London and staged 'one of the first real mass urban parties now known as raves' (Anon. 1993: 9).

> Suffice [it] to say, the idea is to dance as much as you can, do-it-yourself philosophy extending from the making and distribution of the music, through into the setting up of the free-parties by a loose, informal sound-system collective of DJs, sound and light technicians, MCs and general organisers. One of the best examples being Nottingham's DIY sound system who have been organizing parties and providing DJs since 1989.
>
> (Anon. 1993)

During the late 1980s and 1990s, large-scale free parties evolved into 'raves', where 'acid house' and other genres of dance music were played. As noted earlier (see pp. 137–8) at least some participants within rave subculture became politicised after state opposition, with 'ravers' co-operating with EF! (UK) and other radical greens in opposing the Criminal Justice Bill.

DIY practices helped to resource EF! (UK) and the wider anti-roads movement in a number of ways. In particular they 'had picked up some guile from the rave scene's mobilisations', with this contact accelerating the adoption of new technologies of protest like mobile phones and the internet (Collin and Godfrey 1997: 227). DIY culture has come to include diverse forms of cultural reproduction, using increasingly sophisticated mass-marketed items of cultural capital including photocopiers, hand-held videos, small presses and cameras. DIY media has acted as both a supplement and an alternative to the conventional mass media, which is often hostile to radical movements. Activists might take lightweight video cameras on 'actions', producing video cassettes that could be circulated to publicise movement activities and, simultaneously, discourage violence from police or security workers who do not wish to be filmed.

The street party, as noted earlier, developed in part from the rave free-party. Dozens of lorries (an unintended irony) containing sound-systems or occasionally bands to play live music helped attract participants to mass urban road trespasses (Bellos and Vidal 1996: 7). In turn DIY squats have provided office and meeting spaces. Cooltan in Brixton, the centre squatted by Shane Collins, provided office space for both EF! (UK) and the Freedom Network which fought the Criminal Justice Bill. The Rainbow Centre in Kentish Town, a squatted former church, combined accommodation, office space, a cafe and space for cultural activities. It was also used as a meeting place for the first large-scale RTS party in 1995.

While sound-systems, dance and art activities have enhanced the cultural impact of actions such as the street party, contradictions are apparent. Typically, militants in EF! (UK) criticised some of the cultural impact of DIY, seeing it as promoting passively symbolic events to the detriment of disruptive direct action. Such critics feared the transformation of activism into a series of 'spectacles', discouraging activists from indulging in potentially effective if illegal tactics. They also felt that DIY provided an overtly material niche for economically motivated political entrepreneurs, and were highly critical of this eventuality:

> It is a logical progression: people have [squatted] an office, so the media talks to them.... Isn't it much better to take the story to the media or *be* the media even. So that the thing here is small world video and video in general. It is a case where people go around and film everything, [then] they sell it off to the media [and] they make quite a nice living out of it. They present a sort of passive [protest].... There is this 'carhenge', a case of about six people driving cars up to Pollak and burning them and being filmed doing it.... Where is the participation? Where does change come?... Is that any different from Greenpeace paying several thousand pounds and putting an ad in the front page of the *Guardian*, you know, except less people see it?
>
> (Mix in interview)

Equally, as Collin and Godfrey (1997: 227) note, not all ravers have sympathy with the environmental direct action movement:

> [T]he increasing politicisation of the free-party network increased its distance from the club scene, which was becoming ever more mainstream and institutionalised.... [Due to legal sanctions] since 1990 the majority of clubbers had been ensconced in legal venues and focused on drugs, music and fashion, with little common ideology other than hedonism. Although many clubbers perceived the CJB as an attack on their own culture... others felt it was aimed at outlawing parties they never went to and lifestyles they had no intention of following.

Even the goals and assumptions of 'politicised' party-goers may be different from other activists':

> I am not a raver... [and] the CJB... was my first encounter with rave culture. For some people it seems to be quite an unproblematic link: people are quite happy to move between the two identities or not even see them as separate, but I also think that there is a certain amount of antagonism.... At a crude level it is the hedonism of rave versus the kind of value-system of the green movement, and that causes real problems. At Claremont Road it was how tidy you keep the street or how loud you have your rave music playing, and it is quite funny – it occurred at Liverpool again, because we had taken over a

> squat – how clean [we kept], how much we respect, this empty building we were inhabiting, or how much we destroy it, that seemed to work along the rave–eco division....That seemed to be what was going on. The ravers were [saying], 'You just enjoy yourself' and 'You have a good time and you leave litter everywhere', and all the ecos, the serious ecos, are picking up the litter and painting over the graffiti. So there does seem to be all those points where the projects are quite different, and there are problems.
>
> ('Green' in interview)

Lifestyle

An emphasis on cultural practices and 'lifestyle' is common to diverse environmental networks, and many greens argue that cultural strategies are an important means of creating change (Dobson 1990: 139–45; Irvine and Ponton 1988: 138–9; Rainbow 1993: 21–2; Scarce 1990: 25). Dobson notes:

> What seems common to these lifestyle strategies as I have treated them is that they mostly reject the idea that bringing about change is...principally a matter of occupying positions of political power and shifting levers in the right direction. Lifestyle strategies take seriously the idea that profound changes in attitudes are a precondition for social and political change.
>
> (1990: 143)

Linking cultural change and identity formation, Melucci (1989 and 1996) stresses the symbolic challenge that social movements present. He sees such a politics of 'signification' as a means not only of constructing frames for movement mobilisation but of 'reversing the cultural codes' for wider society (1989: 56). A successful movement in this context, while lacking access to the policy process, may come to 'pervade a whole society or culture' (Hall 1985: 291). For many anti-road campaigners, lifestyle change is both an extension of green strategy and a means of survival: economic rationality and movement ethics are seen to coincide. EF! (UK) activists have established housing co-operatives or squats to make a low-cost activist life easier to sustain. Leeds EF! (UK) had an office in Cornerstone Housing Co-op; Manchester EF! (UK) established the Equinox Commune. Activists may live together informally in shared properties, perhaps enjoying communal meals and sub-dividing household tasks. Protest camps are forms of collective living, with all its tensions and its promise.

Interview accounts indicate that an activist's lifestyle may include the rejection of car-use, changes to diet, communal living, the occasional 'ritualised' practice such as the growing of 'dreads' or the avoidance of cosmetic treatments like artificial shampoo, and a general rejection of consumerism. Many activists saw these activities to be politically important in prefiguring a green society, and as a means both of deflecting charges of hypocrisy and of furthering social change.

The culture of the wider anti-roads movement can be more eclectic. For middle-class campaigners, lifestyle change may be alien. For others, living in permanent camps means making compromises in order to survive:

> Food donations have included everything from a bottle of Dom Perignon to 72 tins of EEC Food Aid stewed steak in gravy. The tins have blue labels with a yellow star on, and are basically BSE ridden beef mountain that's been frozen for years. 'Oh no,' says one man on seeing it, 'we had some of that at Pollok. It tastes like a cross between dog food and shit.'
>
> (Merrick 1996: 76)

Communal living, diet, drumming, drugs, dress derived from earlier green cultural networks, all have contributed to the distinctive identity of anti-road camps. Cultural codes, often subtle, were used to maintain boundaries. Thus Styles noted the intensity of personal lifestyle change on entering the movement:

> There were so many things, even though I was a working-class lad, things I had never [thought I] would ever see. [I] slept in a squat for the first time [and] really learnt things that never crossed my path before. Vegan slop was very nice, soya milk was fairly horrible, you know, but that was, like, you live and learn!
>
> (Styles in interview)

Lifestyle does not just generate signs and maintain boundaries: squatting is a means of survival and soya milk an affirmation, in a very practical sense, of support for animal liberation. Yet while formal rules were difficult to implement because of the decentralist ethos, informal cultural codes were used to police EF! (UK):

> Nothing really prevents somebody dodgy from calling themselves an Earth First!er. It is just, you know, if you are leading a mainstream lifestyle, if you are engaging in environmental[ly] destructive activity, if you have rather conventional beliefs, there is nothing to be gained from calling yourself an Earth First!er....Earth First!ers are people who are into living seriously green lifestyles and going around the country doing direct actions.
>
> (Begg in interview)

Some activists, despite long-term participation in the green movement, felt that it was difficult to work with EF! (UK). For example, activists with a working-class identity and a more militant perspective, such as Allen and Garland, came into conflict.

> When I got back there was an Earth First! meeting.... They took me across to the pub and gave me a letter and said: 'Read the letter.' They said I was going my own way, and I said I was almost acting as a mouthpiece for Earth First! but I wasn't, I was doing my own thing. Basically, I fell in conflict, and also I was pushing the ELF stuff quite a bit. And Class War upset them as well. Basically, me and [X], we were outsiders....I was upset, so I said: 'Fuck this. I am going to leave.'
>
> (Garland in interview)

> There doesn't seem to be any process of decision-making that exists at all. It is just a matter of opinion and, um, Earth First! is pretty, like, hermetically sealed, actually, from other groups outside. You very much are in Earth First! or you, you know, don't get much of a look in....Frankly it does raise quite specific problems about who can speak for Earth First!
>
> (Mix in interview)

Thus social exclusion and conflict over issues of repertoire and ideology could become bound together. It is, of course, worth noting that the culture of EF! (UK) may vary from local group to local group; and it has shifted over time (see for example p. 138). Certainly, by 1995 militant sentiments were rather more in favour and key moderates had moved on to new campaigns.

In turn, EF! (UK) activists often situated themselves by reference to the wider anti-roads movement, containing tribalists such as the Dongas. Identities were blurred: some Dongas were hostile towards EF! (UK), while others maintained a dual-identity. Neo-tribalists were often seen as obsessed with a politics of exclusive identity. 'Clare', discussing the Solsbury Hill campaign, argued: 'We let them put all their dogma and all their kind of stuff on us too much....They were somehow a lot wiser...[their] lifestyle was a lot more important than being there to stop the road, and we were all there for *that* reason. We didn't get a lot of respect.'

EF! (UK) activists who rejected the Dongas' culture often recognised its power in building strong affective links:

> They liked doing this Mother Earth kind of stuff, and it did seem to have a real power for all. I had an agnostic view of the world, but this spirituality that went around the camp did have a magical effect, it just did have a real magic. I thought, even if it has nothing more than the bonding effect it is well worth it....I tasted it really powerfully around the camp fire....We would do all these dances around police cars and diggers....It was always good to be doing that to them....I don't know how seriously they took their view, I just liked the effect it had.
>
> (Geffen in interview)

> She did care passionately about the historic aspect of it. We are destroying something [Twyford Down] that has a huge sort of historical, magical importance to a large number of people. That is the point [at which] that druids' stuff came in as well. I remember one action around May [when] loads of druids came. She convinced a lot of people – because she was so passionate – who were dabbling in this stuff and 'the Earth is our Mother!'...Although, having spent twenty years trying to get out of the Catholic thing, I wasn't going to start worshipping something else. That's fine, if that gets you down and on the bulldozers. But I don't think I will get involved, it is not for me.
>
> (Jazz in interview)

Rituals bridging barriers between art, religion and politics imbued action and places of struggle with great affective importance.

> It was just magical. And I will never forget sitting on top of the hill at Twyford, with the lanterns on bushes, you know, thorn bushes, and people sitting around the cooking pots and brewing endless cups of tea and just sitting there talking....Then at five or six o'clock we would have to descend into the valley back to Winchester, and the boring odd sort of roads and the boring station....It was like going...it was like being in a time warp, in a way. It was creating a magical sort of place where people of different aspirations could be.
>
> (Noble in interview)

Such comments illustrate Lofland's belief that 'culture is a double-edged sword'. Intense movement culture may bring 'higher degrees of membership coercion, narrowing of the number and range of people who will participate, and reduction in the civility of the participants' personae and relations among themselves and with outsiders'. Equally it may generate, via the creation of strong identities, a 'greater capacity for collective action, greater tenacity in the face of target resistance and campaign reversals, that derives greater satisfaction from movement participation' (Lofland 1994: 215). Strong culture may also allow movements to create alternatives that genuinely challenge social norms. Thus it has been argued that the Greenham experience created a 'liminal' zone where women could step out of their former routines and re-make identities (Roseneil 1995).

The assumptions of RMT and 'frame realignment' suggested the need to appeal to existing social norms rather than attempting to construct new codes. To influence policy makers and the public, 'frame realignment' would have restricted the Greenham women to a focus on nuclear weapons, to taking tea with colonels wives at Newbury and the masking of their lesbian orientation. Animal liberationists would be restricted to articulating media images and marketing vegetarian foods. Yet without the 'extremism' of EF! (US/UK) or *Green Anarchist*, without the laboratory raids of the ALF and the transgressions of Greenham, it seems unlikely that concerns with animal liberation, sexual alternatives and anti-road struggles would have gained even the modest currency that they have in British society. Outrageous acts, which go beyond the bonds of accepted norms, may have a gravitational pull on society, alienating some individuals but placing formerly unthought of alternatives on the 'agenda'. In short, the challenge is to 'elaborate culture that sustains participation without stifling democratic participation and sponsoring demeaning treatment of non-members' (Lofland 1994: 215).

At worst, EF! (UK) has alienated potential allies with its cultural exclusivity, policing an overtly pluralist organisational form with semi-visible forms of exclusion. At its very best, it has been able to make alliances with environmental pressure groups, local conservationists, marginalised workers and hedonistic counter-culturalists. Culture *is* double-edged: negative aspects cannot easily be identified and separated from the positive. Yet, from the Dongas' fire dances to Twyford Down Association's more conservative symbolism, culture has helped to create and sustain the anti-roads movement.

8 Globalised greens

> Earth First! in Australia never became something in its own right...it remained a kind of rallying cry. You would get a forest blockade, say, run by 'ferals', as they would be called, that are the kind of Donga equivalent...and they would have an Earth First! banner. There was never any Earth First! as such. There were never any Earth First! groups, they never had an Earth First! newsletter. They never had anything like that. But, just as later we did, they took their inspiration from the *Earth First! Journal* in the States.
>
> (Marshall in interview)

Introduction

Earth First! is a powerful example of the globalised nature of modern environmental protest, with groups not only in the US, UK and Australia but in a four-continent matrix of sympathetic networks. While traditional conservation groups, green political parties and environmental SMOs have sought to build sister organisations on a planetary basis (Parkin 1989), EF! (UK)'s international links illustrate how loosely co-ordinated radicals as well as such formal bodies use and extend global links.

The dynamics of diffusion, global and local, are fascinating and mysterious. While it seems that almost 'everything seems to diffuse: rumours, prescription practices, boiled drinking water, totems, hybrid corn, job classification systems, organisational structures, church attendance, national sovereignty', explanations for such widespread diffusion are similarly diverse (Strang and Meyer 1993: 487).

Diffusion can be variously classified. Cross-national diffusion deals with the transference of ideas or symbols from one country to another, as in the movement of EF! from the US to the UK. Intra-movement diffusion describes transferences between groups within a movement. In the case of the anti-roads movement, tactics have been taken from a broader green movement family, with peace and animal-liberation activists contributing to the development of repertoires of action. Inter-movement diffusion is also possible, with distinct and often hostile groups from disparate movements learning from one another.

Rowell (1996) suggests that a 'green backlash' has involved opponents of environmental action using grassroots organising techniques borrowed from their foes to fight EF! and other green radicals. In Britain, conservative countryside campaigners who support hunting have discussed blocking roads and using other repertoires taken from anti-road activists and the animal liberation-movement.

A successful mobilisation may go further and diffuse its essential values to society as a whole. Green strategic success may be measured in terms of diffusion in this widest sense. As Melucci (1996) observes, movements are messengers introducing wide social change by nurturing and transmitting novel cultural codes. Much social movement theory ignores this fourth form of diffusion because of the emphasis it places on movements framing their ideas within what is socially acceptable rather than fundamentally challenging existing codes. Yet societies are transformed by diffusion. Fundamentals such as agriculture, capitalism and the use of writing are all examples of products that have diffused across geographical and cultural barriers to create new social and ecological orders.

Diffusion is thus of considerable political importance. Radical greens might ask how links can be made between campaigners in the northern and southern hemispheres, each with their own cultures, languages and political agendas? Academics are intrigued to know why some campaign issues, ideological features, repertoires of action and types of organisation travel while others do not.

Diffusion theory focuses on differential transference. Much of this theory, particularly when applied to movements, is concerned with networks, culture and political opportunity – themes considered earlier in this book. Diffusion demands either direct channels of personal or institutional contact via networks or indirect influence, usually via the mass media. A wave of student protest in the late 1960s, which apparently erupted simultaneously in Berlin, Paris, Rome and New York, was inspired in part by powerfully iconic news imagery. Yet networks also brought student activists from Europe to the USA and Japan (McAdam and Rucht 1993). Channels, however direct, allow diffusion to occur, but they do not explain why activists or other would-be diffusers feel the need to borrow issues, organisational forms, ideologies, tactics or other products.

Culture, of course, is vital. Diffusion theory emphasises the importance of 'minimum identification': students listen to other students, socialists talk to other socialists. Thus it has been suggested: 'Where actors are seen as falling into the same category, diffusion should be rapid' (Strang and Meyer 1993: 490).

Closer cultural identification clearly makes for an increased willingness to borrow from others; yet, imaginative change may be the product of 'minimal' identification using weak bridges to make new connections. 'Weak' ties may provide 'bridges' into other 'networks' that may provide resources for movement growth; indeed, it has been suggested that '[i]ndividuals with many weak ties are...best placed to diffuse...innovation, since some of those ties will be local bridges' (Granovetter 1973: 1367). Burbridge and Torrance were able to mobilise activists by exploiting relatively weak links with a number of diverse networks. The success of the anti-roads movement both in adapting its tactical repertoire and in mobilising large numbers of protesters has been based on an ability to make links with very diverse social groups from Conservative Party councillors to dock workers.

Kriesi *et al.* emphasise political factors in explaining diffusion, with open political systems allowing waves of protest action to move swiftly from place to place (1995). Clearly, the example of EF!(UK) shows that political closure may motivate activists to seek new repertoires to make their voices heard. Absolute closure and stark repression would, though, make mobilisation difficult.

All three considerations – networks, culture and political opportunity – are significant, but in themselves seem inadequate to account for the occurrence of diffusion in some situations but not others. It seems clear that in the context of the UK green movement cross-national diffusion has accelerated in the 1990s. New conceptual approaches are therefore necessary to explain why this is so.

Environmental problems may produce a particular concern with diffusion. From smog to acid rain to nuclear disaster at Chernobyl to the greenhouse effect and ozone depletion, pollution has become regional, national, continental and, finally, global in scope. Environmental damage may follow international chains of economic production, distribution and consumption. Rainforest timber harvested in Asia travels to shopping outlets in North America and Europe. Increasingly both institutional and radical approaches to the environment have orientated themselves to these global realities.

The instrumental nature of some green diffusion, with environmental movements acting like religions or other political organisations in consciously spreading 'the word', is also important. For example, like EF!, FoE originated in the USA, created when David Brower left the Sierra Club, arguing that its 'preservation approach' was 'too conservative...to match the urgency of what he saw as an emerging ecological crisis' (Weston 1989: 33). Focusing on the globalisation of environmental problems, he sought to create a global movement to combat them. In 1971 he toured Western Europe, and FoE (UK) was created as part of this process of establishing an international organisation. Initially FoE worked with Ballantine Books in Covent Garden to produce a UK edition of the US *Environmental Handbook* (Barr 1971), thus providing the organisation with finance and publicity. Greenpeace International has been even more successful in constructing a planetary identity. In contrast, neither EF! (US) nor EF! (UK) has made an attempt to build a global environmental action empire by deliberately seeking to set up groups on a worldwide basis.

Small militant publications, usually unnoticed by academics or the media, have also acted as informal channels for accelerated international diffusion. *Green*

Anarchist, in particular, has been influential as a vector for new radical influences from North America, Europe and the southern hemisphere. For example, as well as – uniquely at the time – promoting African-American green radicals such as Mumia Abu-Jamal and MOVE (*GA* 1989 (22): 3), it also publicised the US *EF! Journal* and was linked to an early attempt to create an EF! (UK) network (see Chapter 3). The US *EF! Journal* was described in a note of positive identification as 'the most militant environmental paper we've seen' (*GA* Autumn 1987 (17): 16). *Green Anarchist* is an excellent example of Melucci's concept (1996) of a latent or invisible network, apparent in fact to key activists but unknown to policy makers and their researchers.

Earth First! (US) and global links

EF! (US) emerged dramatically out of the experience of its founders in the Mexican desert, and its initial campaign focus was very much on wilderness preservation. While later publications have dealt with global environmental problems, including the greenhouse effect, the very first newsletter advocated the creation within the USA of vast 'ecological preserves' from which 'the developments of man [*sic*] will be obliterated' (Manes 1990: 74). The landscapes of the southern states were EF! (US)'s first love; international green action was to come later – and diffusion to and from EF! (US) came later still.

Its earliest and principal cross-national campaign was rainforest protection, and Mike Roselle, an EF! (US) co-founder, helped create a US-based international Rainforest Action Network (Taylor 1995: 18). While the rainforest has become an icon of international environmental concern (see Chapter 3), rainforests could be linked to EF! (US)'s prioritisation of wilderness in a way that was more difficult to achieve for such global pollution issues as the greenhouse effect. In 1984 EF! (US) co-ordinated demonstrations at more than twenty locations on the rainforest beef issue. This was part of a wider international campaign by different environmental groups against the logging of Central American rainforests to make way for pastures for (fast-food destined) cattle (Lee 1995: 80). The *EF! Journal* has carried numerous stories of rainforest campaigners' resistance. Academic and EF! (US) supporter Bill Devall travelled to Australia to take part in direct action against forest destruction. John Seed, the Australian activist who inspired George Marshall to launch rainforest defence in the UK, had taken part in the 1983 EF! (US) roadshow. According to Zakin, 'the diminutive, fine featured, back-to-the-lander' had made contact with EF! in 1982. His interest was awakened by being shown copies of the *EF! Journal* by deep ecologist and American poet Gary Snyder (Taylor 1995: 18). By the mid-1980s Seed was contributing a regular column called 'Letter from Australia' to the *EF! Journal*.

In 1984 Australian and Swedish correspondents were regularly contributing to the pages of the august 'Shit Fer Brains' letters' section of the *EF! Journal*. At this point only three international affiliates were reported in the journal's list of contacts. A travelling American in Kyoto, Japan, and two Australian Rainforest Campaigners hardly provide evidence of effective cross-national diffusion, whether to or from the US (Lee 1995: 174).

By the late 1980s international contacts had grown to include EF! groups in Australia, Canada and Mexico, and Chris Laughton in Britain. The nature of Laughton's activity, which as noted in Chapter 3 (pp. 45–6) failed to sustain even a cell of EF! supporters in the UK, suggests that these contacts do not necessarily provide evidence of thriving radical international action. Nonetheless EF! was making tentative green network links outside of North America.

EF! Australia was composed largely of 'ferals' , or travelling activists with no fixed affiliation, who established pro-rainforest blockades. Intriguingly, a study of the movement entitled simply *Earth First* contains no addresses for members of the network in its extensive list of rainforest campaigning groups in Australia. Repertoires of action from Australia have been seen as inspiring tree-sitting in EF! (US) protests (Zakin 1993: 250). At least one author has suggested that Australian pro-rainforest activists were far more successful in promoting mass direct action protests than was EF! (US). Zakin's comment (1993: 249) that Australian protests were 'far larger than...[EF! (US)] could even dream about' (as many as 1,500 campaigners attended), suggest that the North American movement at the time was tiny.

The EF! (US) of the late 1990s seems to be fractured as between young anarchist militants similar to many EF! (UK) or European EF! activists (see below), exponents of strict Gandhian non-violence and the original Foremanites. The young anarchists are represented by *Live Wild or Die* and the Foremanite deep ecologists by *Wild Earth*. Diffusion has brought in new groups of campaigners, and Toker (1997: 140) notes that in 1996 EF! (US) established a 'Campaign to End Corporate Dominance' to link together labour activists, native groups and social-justice campaigners. EF! (US) has been involved in a modest anti-roads movement with a number of other groups, and has sponsored the journal *Auto-Free Times*. Typical anti-road actions have focused on new routes through wilderness areas. Such actions have mobilised relatively small numbers; mass direct action featuring thousands of local conservationists and youth counter-culturalists has been notably absent. One account of a 1993 anti-road camp in Dixie, Idaho, suggests a context of high polarisation:

> The townsfolk saw themselves as persecuted. The actions of Earth First! directly threatened their livelihood, they claimed. A cardboard sign propped against a roadside gas pump warned, 'Earth First! Don't Stop,' and a notice taped to a window of the Lodgepole Pine Inn warned: 'We reserve the right to refuse service to all Earth First!ers and their associates.' A poster at one end of a stringy-haired Earth First! activist with a bullet in his

> forehead....There was evidence, denied by Earth First!, of traditional monkey-wrenching tactics....There were also signs that the locals were taking out their frustrations in more hostile ways. One afternoon a gang of drunken men ambushed one activist on a lonely road and beat him bloody.
>
> (Kine 1997: 128)

In the UK violence has been used against EF! and counter-movements supporting targeted bypasses have emerged, yet the anti-road campaigns have been built around local support networks. Often, as at Twyford, local communities have invited EF! (UK) to establish campaigns. Even given the bias of many academic and media accounts hostile to EF! (US), the contrast is clear.

Many EF! (US) groups seem to practise surprisingly non-confrontational and ritualistic demonstrations, partly in fear of repression. Zisk (1992) noted that EF! San Francisco took part in largely conventional demonstrations. An article by EF! (US) in *Do or Die!* (1997) provides a strong contrast with the radical sentiments of the UK anti-roads movement and EF! (UK). The campaign against redwood-tree destruction made an appeal for individuals to take 'a wrench to the bureaucratic wheels by attending agency hearings, reviewing logging plans, and drafting initiatives to reform California forest policy' (1997 (6): 127). Physical monkey-wrenching is rejected on practical terms: 'By refusing to damage equipment or harass loggers, we have reached a fragile, unspoken state of affairs where very few activists are hurt – except by the cops' (*Do or Die!* 1997 (6): 125). Such statements, even if they represent only a fraction of EF! (US) activism, hint at a movement that is somewhat weak within a strong and hostile political system. Thus there is some evidence to suggest that EF! (US), to the extent that it inspired EF! (UK) and EF! in Australia, was the weak parent of rather stronger children. Despite the mythology that has accrued around EF! (US), EF! (US)'s enduring legacy, from Abbey's anti-humanism to the bombing of the late eco-socialist Bari, may be the diffusion of such dramatic tales as an impetus to militant global green action.

EF! Australia

The tunnelling tactics used on the A30 and at Manchester Airport protests have their origin in the rainforests of New South Wales. Marshall had travelled to Australia and returned to the UK committed to creating a direct action environmental movement at the grassroots level. Activists with Australian experience, such as Shelley Braithwaite, had inspired EF! (UK)'s founders, Burbridge and Torrance. Begg revealed how an Australian pamphlet, entitled 'The inter-continental blockaders' guide', had directly influenced tactics used at the M11 and M65 anti-road actions.

Protest against the flooding of Lake Pedder National Park in Tasmania began in 1965. The world's first ecological political party contested elections in the state in the early 1970s (Parkin 1989). In turn, the first time that 'a large group of people were defending a forest physically by standing between the bulldozers and the trees' is reported to have occurred in the Terania Forest of New South Wales in 1979. The area was a magnet for counter-culturalists, who had established communal villages there, as well as an Aquarius Festival in 1973. Three hundred activists halted logging being done by the Forestry Commission and established a protest camp. The campaign was ended by an act of very public tree-spiking, when a 19-year-old student climbed a tree, inserted a number of nails and proclaimed: 'The bastards won't be milling or cutting down this lot' (Kendell and Buivids 1987: 47). An Australian Broadcasting Corporation account noted:

> This act of vandalism, although disowned by the protesters, effectively stopped logging. It became clear that massive police protection would be needed not only to continue the logging but also to prevent spiking and destruction of felled trees.
>
> (Quoted in Kendell and Buivids 1987: 55)

Tasmania, December 1982, saw the arrest of 700 protesters using blockades to halt construction of the Franklin Dam. The strictly non-violent direct action was based on training developed by the US Movement for a New Society:

> Its aim is to build affinity groups of about 10 people – groups that will be capable of making decisions about their tactics, camp organisation, and policy – and also to develop the internal awareness of the group, to help solidarity, caring and mutual support. The training involves presenting groups with certain tools – techniques to develop consensus decision making and group dynamics, conflict resolution, relaxation exercises and a general overview of non-violence theory.
>
> (Paasonen 1983: 13)

The parallels in this account with Greenham Common, the US peace and civil rights movements and EF! (UK) are striking. Indeed, the activists were inspired and informed by the experience of American anti-nuclear protesters, who, in turn, used repertoires diffused from the civil rights struggles of the 1960s (McAdam 1982). The civil rights' movement in turn had been inspired by Gandhian methods used in India and South Africa.

Direct action was also used in an attempt to halt road construction through the Daintree area of rainforest in Northern Queensland in 1983. Activists described Daintree as 'the last remaining...virgin rainforest left in Australia'. Home to the unique Bennet's Tree Kangaroo, 10 per cent of the species in Daintree were said to have been unrecorded when construction began. Repertoires used in Daintree diffused to both EF! (US) and, ultimately, EF! (UK) and the anti-roads movement. Tactics were based on placing activists in situations of deliberately constructed danger and were the precursor of the tunnelling techniques used by UK road protesters.

In interview Marshall noted:

> I couldn't become that involved in direct actions there because I was considering at that time actually staying in Australia, and I did not want any kind of a criminal record. You know what it is like if you get arrested and your name is on some bloody computer somewhere. I didn't even want that, although I was involved in forest protests. I ran the risk of being arrested but I didn't put myself in the front line to be arrested. People were using tactics that it took years for people to start doing here in Britain, like burying themselves in the road chained to blocks of concrete. All of this kind of thing was happening in Australia. . . . I would be the one who would dig the hole and pour the concrete.

Whereas radical green activists in the UK sought a new slogan and a new focus from North America so as to resist the institutionalisation of the green movement, John Seed was inspired by feelings of international solidarity. Whereas EF! (UK) acted as a catalyst for anti-roads direct action, EF! Australia fitted into an existing mobilisation:

> [T]he Australian Rainforest Campaign started with direct action right at the very beginning with probably the very first full-on direct action campaign to save an area of wilderness or an area of the environment in the, shall we say, the developed world, the non-third world, the first world, which was the Terrine Creek process in northern New South Wales in the late 1970s. That was something that was very much coming from the new age movement in northern New South Wales. That experience of direct action, that experience of people who had settled up there fighting to save something they felt passionately about had created a whole energy within the rainforest movement, and produced a whole lot of activists who then spilled out. Around about the same time as that, and also very much inspired by that, had been the Franklin River blockades and campaigns down in Tasmania.
>
> (Marshall in interview)

Thus, unlike in the genesis of EF! in the US or the UK, political closure does not seem to have been an important influence on the emergence of EF! Australia. Ideology there seems to have been a stronger link than in the UK, with Seed advocating deep-ecology views, including a misanthropic parable delivered to the (US) *EF! Journal* in the form of 'An immodest proposal'. In the tradition of Manes' anti-humanism, rather than Swift's satire of Malthus that suggested a diet of Irish pauper infants, Seed praised the benefits of a mythical virus that kills human beings but leaves other species and wild nature unscathed. Seed has also stressed his support for ecotage, quoting Edward Abbey and noting that even Gandhi preferred controversial forms of action to non-resistance (Kendell and Buivids 1987).

In contrast, Marshall noted:

> The Rainforest action groups were practising a strict, and sometimes in my view over-strict, Gandhian NVDA. Over-strict to the extent that they would get into intense debates about whether to break a lock on a barrier, whether to cut through a wire fence. This always seemed to me [to be] crap. You know, who the hell really cares.... They were, however, running a very high media campaign to win the hearts and minds of people, and they had had unpleasant experiences in the past where, you know, the media had taken advantage of somebody, for example, slashing the seat on a bulldozer, which was very minor criminal damage, to smear the campaign. So they were very wary of that.
>
> (Interview)

Several features of EF! Australia are shared with EF! (UK). In particular, EF! Australia has a minimal institutional existence and is a network within a much larger non-violent direct action movement. Anti-logging blockades are magnets for 'ferals' or counter-culturalists similar to young anti-road activists in the UK.

Diffusion from the UK to Australia has also occurred. Seeking wilderness, Jason Torrance moved to Tasmania in 1995. Reclaim the Streets actions have also spread there. In November 1997 the first RTS was held in Sydney when the busy Enmore Street was blocked by three giant bamboo tripods: 'A sound system in a huge art-installation tower on wheels rolled into the street, and a permaculture garden was constructed in the middle of the road' (*Peace News* 1997 (2420): 5). The event was organised with the reported co-operation of the police, and local firms, with the exception of petrol stations, were reported to have increased their business. It is difficult to imagine an RTS action in the UK promoting business and being empowered by police protection.

Earth First! Ireland

Robert Allen was involved in the creation of a modest Irish EF! network, as well as being active in Britain. Like EF! (UK) the Irish movement, An Talamh Glas (Gaelic for Green Earth), is a loose network which makes use of direct action against road construction, sustaining its identity with an occasional action-update newsletter. Davy Garland claimed to have undertaken EF! action in Northern Ireland prior to the creation of EF! (UK), and a student-based Belfast EF! group was created in the early 1990s.

In 1997 activists, including Allen as well as individuals from the UK and North America, helped to initiate an anti-road camp in County Wicklow with the intention of preventing the widening of the N11. EF! (UK)'s *Action Update* encouraged activists to support the protest:

> Glen of the Down is as good as it sounds and it's about to be destroyed unless we get people there to stop the Irish state using EU (and British) money to expand Ireland's bit of the Trans European Road network....Known as the Garden of Ireland, County Wicklow, with its golden mountains, hidden valleys, gleaming lakes, gushing rivers and flowery forests, does not need a wider road. The campaign has suddenly become news among Ireland's sensationalist media even though the bulldozers are still silent, so people with ecological road sensibilities are desperately needed.

Campers constructed tree platforms similar to those used in the UK's M65 campaign, and activists from Newbury and Scottish road protesters were involved. The local *Wicklow Times* noted: 'They have come from all around the world for this protest and at present number around 10 to 15. "I used to be a computer software designer," said one of the warriors from California, who called himself "Clinton".'

The project, funded with £20 million of European Community regional aid, aimed to create a four-lane motorway, which would be part of a route stretching from Rosslare to Belfast. Large sums of money have been transferred from EC regional funds into Irish road construction as part of a larger project of industrial development, which has helped fuel the high economic growth-rates and rapid modernisation in the country. The protesters at the Glen of the Downs camp feared that road-widening would lead to the felling of 2,000 mature ash, oak and beech trees together with the devastation of an area supporting much wildlife, including sika deer and red squirrels.

The Irish media regarded the camp as a direct extension of British roads protest. In turn, some groups within the well-established Irish green movement were critical of what they perceived to be 'foreign' intervention. Nuala Ahern, an Irish Green Party MEP who lived near the camp, attacked the protesters, observing: 'I would hate people to get the impression that we need foreigners to protect our trees' (*Wicklow Times* 24 September 1997: 1). The Irish Green Party had led a long-term protest against the widening of the road, but felt that significant concessions had been won in that the proposed width of the N11 had been reduced.

While fast economic growth, apparent in the 1990s, is likely to fuel environmental damage in Ireland, proportional representation has provided a more fertile and open political opportunity structure for the green movement. In contrast to the UK, Irish governments are relatively short-lived, being constructed around coalitions of small parties. Although the Green Party has not to date (late 1998) participated in government, it has some influence on policy matters along with representation at the European level. The former Irish Environment Minister Liz McManus is reported to have opposed the N11 project (*Wicklow Times* 24 September 1997: 1). Equally, the swift growth of the Irish economy during the 1990s may have strengthened public support for road construction and other environmentally destructive capital projects that deliver apparent prosperity.

Other forms of EF! action have diffused to Ireland. The November 1997 EF! (UK) *Action Update* reported that the Gaelic Earth Liberation Front had claimed to have destroyed Ireland's first genetically engineered crop. The crop of sugar beet planted under an Irish Environmental Protection Agency licence near Carlow as part of a US-based corporation's experiment was uprooted. Strong contact with activists in the US and the UK, and the influx of young environment-conscious Continental Europeans have facilitated diffusion, but policy openness may limit the growth of non-violent direct action. The hostility of some elements within the Irish Green Party mirrors the initial opposition of environmental pressure groups in the UK, such as that of FoE to the emerging EF! (UK).

Earth First! Germany

Germany has a lengthy history of militant environmental action. Moreover, a tendency is evident to link green politics to the German formal policy process. Together, these factors would seem to count against the creation of an EF! network in the country. As noted in Chapter 6, the environmental movement is assisted by the German state, and so discouraged from militancy, while green militants are strongly repressed by the state. Equally, to the extent that green activists practise repertoires of either sabotage or mass action, such repertoires already exist in Germany and there seems little need for them to be imported from the UK or North America. Nonetheless a small number of EF! groups exist in Germany, and anti-road actions have taken place. Such groups have been seen by German activists as a product of diffusion from the UK: 'Some – mainly young – activists went on to form Earth First! groups, most of them in 1994, in many cases inspired by Earth First! in Britain (so you can be proud of that!)' (*Do or Die!* 1997 (6): 109). Activists from EF! (UK) toured Germany in May 1997. According to the April 1997 *Action Update*:

> The NVDA tour to Germany is happening this May. People from different groups involved in NVDA here in Britain will be sharing skills and experiences.... We're visiting groups involved in genetics, roads, anti-fascism, peace and absolutely everything.

The initial German EF! groups were said to have been inspired by German frontline punks who mimicked right-wing US equivalents. Green socialists have accused German EF!ers of advocating a far-right ideology. In particular, Jutta Ditfurth, the left-wing former Die Gruenen activist, seems to have been inspired in part by Murray Bookchin's polemics against EF! (US). *Do or Die!* (1997 (6): 110) noted:

> In her new book *Entspannt in die Barbarei* (*Relaxed into Barbarism*) she speaks of fascist biocentrism, fanatic vegans and the construed connections of radical ecologists to fascists. In Gorleben, when she came to congratulate four activists who had stopped the Castor by climbing up trees and put up a walkway over the road, the reaction of the crowd was to ironically yell: 'Hey, somebody get those ecofascists out of the trees!' Poor Jutta was received with boos and hisses and she lost her composure and stuck her middle finger up to everyone.

Both German and British EF! activists have participated in mass direct action against nuclear power as part of a tradition stretching back to the 1970s that diffused repertoires to the US anti-nuclear protests. A number of anti-road camps have also been established. 'Anatopia', a protest camp established in 1990 to prevent the construction of a Mercedes testing area, survived until 1995.

At Freiberg, fifty activists established a temporary anti-motorway tree-top camp by fixing between four trees a net that had been used in the east London M11 protest at Claremont Road (*Do or Die!* 1997 (6): 115). The campaign was the culmination of twenty-three years of opposition to the construction of the 'B31 East' (*Peace News* 1997 (2408): 7). The camp was well resourced:

> [W]e had everything we wanted in the trees. On one platform, we had a computer with internet and fax powered by solar...another platform was for the media; we would lower it down and hoist them up with it. Rampenplan gave us a kitchen for a ground support camp. All the platforms were six by eight foot [*sic*] with tarps fixed like a tent and swinging from branches.
>
> (*Do or Die!* 1997 (6): 112)

Despite being tipped off via the internet – apparently by the mother of a local police officer – about the date of the planned eviction of their camp, and in the face of protest from several hundred local supporters, 400 police officers removed the campaigners after three days of tree-top protest in December 1996. Forty campaigners were moved from the trees, while 400 other protesters attempted to mount a blockade (*Peace News* 1997 (2408): 7). The protesters blocked the streets of Freiberg in a further anti-road action, bringing together all kinds of people, 'young, old, anarchists and doctors all protesting against pollution, cars and traffic' (ibid.).

In the former East Germany, protests against the Erfurt–Coburg motorway project were launched in 1996 when 'a handful of powerful activists started to initiate an as yet unknown kind of resistance, the squatting of trees to save them from the stupidity of politics personified in chainsaws' . The 'Bettelmanns Holz' were part of 'the biggest still existing forests in Germany, and were expected to give way for the most expensive and destructive motorway in our history' (*Do or Die!* 1997 (6): 110). Many of the protesters were vegan, as were EF! (UK) activists, and so were pleasantly surprised to be cooked vegan meals by local residents. Locals also supported mass non-violent protests: 'a forbidden demonstration under the beeches took place because a local priest declared it a mass(!) and so the hundred cops with their dogs weren't able to interfere'; ecotage was also carried out, with 'the monkeywrenching of several diggers' (*Do or Die!* 1997 (6): 110). Eventually thick snowfalls diminished police attempts at eviction.

Anti-road camps were also created in northern Germany. The Greifswald camp was established in the mid-1990s, but has been a victim of assaults by far-right groups. An internet appeal for solidarity in 1997 noted that fascists had attacked an anti-road camp:

> On Friday the 28th March 1997 the camp against the A20 road construction site, known as the Baltic Sea motorway, was attacked by...approximately 40 right-wing extremists [using] baseball bats...two camp occupiers were seriously injured and one other was badly hurt.
>
> (Free Information Network, 9 April 1997)

German EF! is apparently a small player within larger extant militant networks and a green movement practising conventional political repertoires. Road camps have attracted militant anti-fascists, anarchists and supporters of environmental direct action. Militant environmental action, including sabotage and permanent camps, has a lengthy history in the country. Road protest, too, has been a campaign focus in Germany for decades, though contact with EF! (UK) has contributed to a small number of German activists borrowing UK repertoires which have bolstered their anti-road actions. EF! Germany, though, does not seem to have been driven by the same imperatives as have EF! (US) or EF! (UK). Cultural identification and personal contact with British activists rather than POS changes seem to have been the diffusion factors important to its organisation and modest activities.

France

Burbridge, Torrance and Chris Lang, the latter an Oxford EF! activist, visited the Vale of the Aspe protest camp in 1993 to support the long-standing campaign against a road through the Pyrenees. Local campaigners from the Coordination pour la Sauvegarde Active de la Vallee d' Aspe (CASVA) were joined by the British EF!ers, plus activists from Belgium, Germany and Holland mobilised by Youth and Environmental Europe. Actions were similar to those seen in the UK, but the official response was swift and brutal:

> Actions from the camp included a 20km march to the construction site leading to the occupation of the site including blockades of machinery resulting in work being temporarily stopped; blockades of roads; demonstrations outside the offices of DDE (the government department responsible for the works); and demonstrations, leafleting and singing and dancing in the local towns.
>
> Despite the peaceful nature of all the demonstrations, the police response was 200 riot police using rubber bullets, batons and CS gas...
>
> (Lang 1992: 19)

Two years later, *Action Update* was still appealing for activists to travel to France:

> Anyone in need of NVDA on foreign shores would do well to hitch to the Valley of the Aspe, South West of Toulouse, France, where an anti-road camp is established, but as usual always in need of more people. The mountain air would do you the world of good.
>
> (*AU* 1995 (13): 3)

A modest French EF! network has existed since the early 1990s. Despite fierce repression of direct actionists and the maintaining of a strong civil and military nuclear programme in the 1970s and 1980s, elections in 1997 saw the creation of a coalition Socialist government with Green Party support and the shelving of some environmentally damaging capital projects.

Dutch Earth First!

Groen Front!, which describes itself as 'the Dutch EF!' , was created in 1996, and has initiated a number of anti-car actions. Its first action, intended to draw attention to the contribution to global warming of air travel, saw the disruption of a Dutch television programme promoting tourism. Four weeks later, nine activists occupied the site of a planned road bridge, before being evicted. While the numbers involved in these events, according to activists' own accounts, can be measured by the digits of two hands, a 1997 EF! (UK)-supported Reclaim the Streets party in Amsterdam drew around 1,000 participants.

Prior to the creation of Groen Front!, a number of Earth Liberation Actions (ELF) were claimed in Holland (*AU* 1993 (8): 2). A number of the EF! (UK) interviewees who are sympathetic to ecotage enthused over the actions of the Dutch ELF:

> [19]93 was the best really....All hell broke loose again, the ELF were running around Europe, networking with all manner of things. They were basically, you know, laying down the roots of more resistance and Earth Nights. October Halloween was the best ever. I understand [that] in Holland alone...fifty machines were knocked out. The Animal Liberation Front came out and knocked out lots of hunting equipment and that [they were] also hitting vivisection laboratories, [and] cars...including the Mayor's car in Leiden. Yeah, shops were hit. It got ten minutes publicity on television.
>
> (Garland in interview)

Both in Holland and Germany anarchist autonomists and animal liberation activists supported ELF 'Earth Nights' publicised from the UK. Claiming instrumental diffusion, a Dutch newspaper trumpeted 'British Eco-terrorists cross the North Sea' . In response *Do or Die!* (1994 (4): 37) noted: '[T]he Dutch establishment obviously can't stomach the idea of a radical ecological movement rising up from its own country, so it has to be, of course, the work of outsiders.' Before EF! (UK) or ELF were created, at least one Dutch activist had embarked on a strategy of burning bulldozers: Paul S. undertook seventeen 'eco-sabotage actions' before being caught following the showing of a video of one such attack on a television crime show.

Eastern Europe

Do or Die! contains numerous reports from the Czech Republic, Poland, Slovakia and the former Soviet Union. EF! groups exist in Poland, and Ecodefense, modelled in part on EF!, is active in Russia. All were created in response to serious environmental problems. Nuclear waste and chemical dumps are legacies of the former 'command economies', while pressure for free-market development has led to accelerated pressure on habitats.

The Russian group Ecodefense is committed to 'principles of deep ecology and biocentrism, combined with social responsibility and social justice' (*Do or Die!* 1997 (6): 128). Action camps have been successful in halting a number of destructive projects, including the closure of a large chemical plant in the Ukraine: non-violent direct action has been combined with parliamentary lobbying and educational work.

The Polish Green Brigades-EF!, which 'identify with Earth First!, [using] the same symbols and concepts of activity', have been active since 1996 (*Do or Die!* 1997 (6): 101). Since the early 1990s, Pracownia Na Rzcz (the Workshop for All Beings) has actively pursued deep-ecology projects in defence of the Polish wilderness, while the larger Green Federation has been active in opposing motorway construction. *Action Update* (1998 (49): 2) reports that Poland EF! has being active building tree-houses in the path of the A4 motorway, noting that 'since this is the first protest of this kind in Poland, the authorities don't know what to do, and there's been a stalemate ever since'.

Despite the fact that the command economies have collapsed only recently, East European countries, too, have long traditions of green concern. Green movements were highly active during the events leading up to the collapse, and their perceptions of wilderness support are congruent with deep-ecology concepts. Direct action environmental groups have made contact with EF! (UK) but, as in western Continental Europe, EF! groups comprise only a small element of much larger green networks which include Green Parties, FoE branches, environmental SMOs, anarchist groups, peace campaigners and animal liberationists.

Earth First! elsewhere

EF! contacts have been listed for Egypt, India and Mexico by the US *EF! Journal* and the UK's *Action Update*. An active South African group, Young Lions EF!, was created in 1996. EF! (UK)'s *Action Update* (1996 (29): 4) reported that a twelve-strong roadshow had toured South African townships as well as towns in Botswana, Mozambique and Zimbabwe. A public-address system, an EF! flag and 1,500 pamphlets translated into Zeus, Zulu and Sotho accompanied the activists.

> They drove to townships and played music and displayed the flag to draw attention to themselves, and soon they had a lot of young people at the kombi enjoying music and dancing. They then talked to them about their immediate environment – shacks, garbage, water pollution and trees cut down for firewood. Ignore all the stories about crime, rape and pillage that the papers are stuffed with. They said that people were friendly and wherever they went they were offered food and a place to stay as well as money to help fund the rest of the trip.

Canadian EF! groups have campaigned on forest issues and have established links with EF! (UK). EF! has been active in opposing dam construction in Quebec (Toker 1997), and US EF! groups have lobbied to prevent electricity from the Canadian projects being used by US states.

Solidarity and struggle

Relatively few of EF! (UK)'s international contacts are with named EF! groups or in networks such as Russia's Ecodefense which in part model themselves on EF!, in either its US or its UK manifestation. Most of EF! (UK)'s international contacts are based in direct action campaigning groups which have a shared issue-focus. Beyond anti-road campaigns, many EF! (UK) actions focus on international solidarity work. EF! (UK)'s international solidarity actions tend to support grassroot radical environmentalists, particularly those who advocate militant tactics and who may be otherwise ignored by UK environmental campaigners. Typically, such grassroot radicals may be subject to considerable repression, especially if campaigning on issues with a clear UK involvement. The Bougainville protest is typical of EF! (UK) solidarity actions.

Bougainville is a large island to the north of Papua New Guinea (PNG). While officially a part of New Guinea, many of its inhabitants express an affinity with the Solomon Islands and have supported moves to gain independence. A huge copper-mine on the island, partly owned by a subsidiary of Rio Tinto Zinc, provided much of PNG's export earnings in the 1980s. A mine manager, according to Quodling (1991: 1), has admitted that 'resultant disposal of waste and tailing materials assume massive physical proportions'. Simmering conflict over the independence issue and concern that the mine was poisoning the island led to the creation, in 1988, of the Bougainville Revolutionary Army (BRA), a guerrilla group. Led by Francis Ona, a former mining surveyor, the group assassinated a number of mine employees and blow up power lines. In response, the mine was closed and the BRA gained control over the island. PNG mounted a blockade of Bougainville, cutting off both medical and food supplies, fuelling a number of epidemics. Amnesty International has documented evidence of considerable human rights abuse by PNG forces. Attempts to recruit a British-based mercenary force led to a political crisis and the collapse of the PNG government.

The PNGs living near the mine have a strong land-ethic and reject authority.

> Early German administrators reported difficulty in finding local leaders through whom to exercise colonial authority. The Australian administration fostered broader political awareness, but the grass roots village society was characterised by an absence of structured authority.
>
> (Quodling 1991: 11)

Quodling, in an account strongly biased towards the mine-owners, quoted Bougainville students in the 1970s appalled by the seizure of land for the gigantic mine: 'Land is our life, land is our physical life – food and sustenance. Land is our social life; it is marriage; it is status; it is security; it is politics; in fact, it is our only world. When you take our land, you cut away the very heart of our existence' (Quodling 1991: 11).

EF! (UK), sympathetic to such views, has attempted to support Bougainville independence. Contacts with representatives of the Bougainville authorities in the Solomon Islands and Australian Bougainville Solidarity campaigners in the UK have led to EF! (UK) actions to embarrass the Australian government and raise the profile of the campaign.

The April 1997 *Action Update* observed:

> These people are involved in probably the first environmental war with hundred thousand of them in concentration camps, need our help. They're against the evilest earth destroyers we have ever seen, RTZ/CRA executives helped by Sand Line and the P.N.G. armed forces.

EF! (UK)'s first international solidarity campaign was in support of the Malaysian Penan's campaign to preserve their rainforests from the efforts of logging (see pp. 50–2). EF! (US) contacts brought together the EF! (UK) activists, members of the German Robins' Wood organisation and Australian rainforest activists. Solidarity actions in the UK included disruptive demonstrations at the Malaysian airline company and the Malaysian Embassy.

While the Nigerian Ogoni, in contrast to the BRA, have been supported by larger, more respectable, organisations including Greenpeace International and the Body Shop, they have attracted strong EF! (UK) solidarity, too. Shell, which has been accused of devastating Ogoniland during oil production, have seen EF! (UK) actions against their petrol outlets and the disruption of company meetings.

> Earth First! sees the link between the Ogoni fight for self-determination and the social disillusionment growing in grass-roots Britain. Some feel that the repression of the Ogonis by the Nigerian authorities has echoes of British Home Secretary Michael Howard's crusade against travellers, protesters and animal rights activists; but also they can see [that] the oil which Shell has been pumping out of Ogoni land fuels the world's cars and the inevitable destruction and pollution of the countryside.
>
> Nine of them rush the police standing outside the High Commission and padlock themselves to the railings. In minutes they have been hauled away and charged for trespassing on diplomatic territory. They read out the names of the nine Ogonis ordered to die in Port Harcourt.
>
> (Vidal 1997: 267–8)

Support has also been given to the radical MOVE organisation based in Philadelphia (see p. 7). I initiated contact between EF! (UK) and MOVE in 1994 by helping to write a *Do or Die!* article.

A picket and protest camp organised outside the US Embassy in 1995, when MOVE supporter Mumia Abu-Jamal was threatened with execution, was supported by EF! (UK), branches of which organised a number of other actions in support of MOVE: South Downs EF!, the militant Brighton-based group, occupied the roof of the company American Express; Avon Gorge EF! activists symbolically burnt a national flag of the USA and a mock electric-chair in Bristol city-centre. After Mumia was given an indefinite stay of execution, a protest by Anarchist Black Cross, a prisoner-support group, attracted EF! (UK) activists to an occupation of the Regents Street Disney Store in London. While forming only one element of the wider campaign, EF! (UK) activists spearheaded direct action protests in solidarity with MOVE and Mumia. Shared beliefs and revulsion at the violence meted out to MOVE meant that a MOVE tour in 1996, when Ramona, Carlos and Sue Africa spoke, was strongly supported by EF! (UK). The MOVE activists also visited the Newbury protest during their stay. In the US, MOVE forged a friendship with radical EF! activists, such as the late Judi Bari, and the two networks remain in touch (personal communication to the author from Ramona Africa).

The dialectic of globalisation

EF! (UK)'s international campaigns focused initially on linking environmental destruction in the southern hemisphere to chains of consequences leading back, ultimately, to the UK. Typically, timber cut from the forests of Sarawak would be traced to shipments to the UK and port blockades would be initiated. Forest timber depots and shops selling the timber were also picketed. In turn, EF! (UK) has moved from particular international environmental actions to a generalised revolt against globalisation. Globalisation is a trend towards global economic and political activity, with flows of capital, labour, political power and information transcending local boundaries. The tendency towards the creation of a single global free market built on the basis of regional trading blocks, including the European Union and NAFTA and cemented by the Urguary round of the GATT agreement, fuels the environmentally damaging and socially dislocating economic growth that EF! (UK) so strongly rejects. Most dramatically, the network has shown sympathy for the Zapatista rebellion in southern Mexico, launched on 1 January 1995, against the foundation of the North American Free Trade Agreement (NAFTA). Campaigns against genetic engineering and air travel may also be seen as part of this revolt.

Globalisation, while a target of green protest, has also led to diffusion between EF! (UK) and international groups. Capitalist accumulation, by creating 'time–space compression', via new and ever-more efficient means of communication and transportation, also accelerates the rate at which repertoires can be learnt and solidarity networks established (Harvey 1990: 147). Thus time–space compression has made it easier for movements in the 1990s to grow and hybridise. In the late nineteenth century, British radicals with green political concerns, such as William Morris, came into contact with the writings of such political exiles as Marx and Kropotkin (Gould 1988: 16). Working-class environmental concern was equally influenced by diffusion from North American literary sources, with individuals establishing Whitman Clubs based on the thoughts of the nature poet (Gould 1988: 32). Travel remained slow and diffusion occurred over periods of years or even decades. By the 1960s information was transmitted via the mass-media in minutes, and air travel facilitated instrumental diffusion, such as the creation of FoE and Greenpeace in the 1970s. Television and transatlantic flights helped accelerate such diffusion.

The anti-roads movement could, by the 1990s, call upon myriad forms of movement technology such as the internet, the mobile phone and the camcorder. All were products of capitalist accumulation, which increased the speed of innovation, 'globalised' environmental struggles and made protest mobilisations easier: 'Mobile phones and CBs are great. God, all this would've been so much harder ten or twenty years ago' (Merrick 1996: 10).

Thus, using the internet, activists in a tree in Germany or a tunnel in the north of England can appeal for solidarity to like-minded individuals in North America or Southern Africa. Global economic accumulation fuels grievance by prompting the construction of more airports, roads and high-speed rail-links, but paradoxically enables activists also to mobilise more effectively than ever before. EF! (UK) originated in response not only to perceived political closure in the UK but as part of a large and growing wave of concern over the consequences of global environmental damage. It seeks to build resistance on a global scale. The shift from local environments in the southern USA to such planetary green politics is striking.

9 The preferred way of doing things

> To a large extent it needed someone to stand up and say Earth First! has started, and the Hastings group did that. And from there on the idea of Earth First! was bound to arouse a lot of interest, because people who knew about it [had been] thinking for some time: '*This* is what we ought to be doing here, Friends of the Earth and Greenpeace [are] not radical enough.' This was especially true in the wake of 1989 and the run-up to the Rio Summit.... There was quite a bit of demand for [FoE and Greenpeace's] services as lobbyists...at these big international events, and as a result they were neglecting the high-powered campaigning...So there was this gap in terms of radical campaigning.
>
> (Begg in interview)

Strategy and success?

The direct action anti-roads movement of the 1990s was given a kick-start by green activists frustrated and dismayed by the ineffectiveness of conventional environmental groups and traditional lobbying techniques. There is some evidence to suggest that such 'radical campaigning' has shifted the balance of power between pro- and anti-road advocates. Noting the British Roads' Federation's strong links with the DoT, not to mention a budget nine times larger than that available to public-transport lobbyists –'in a world where influence can be gained over lunch, a fat expense account can buy an awful lot of lunches...to flatter relatively poorly paid targets, such as civil servants and MPs' – Hamer's account (1987: 131) of the pro-roads lobby in the period 1960–89 remains a pessimistic hymn to instrumental power. The anti-roads lobby was portrayed as being ill-funded, poorly co-ordinated and uninfluential, 'like a first-division football team playing a non-league side' (Hamer 1987: 129). During the 1990s a financially and technologically well-endowed pro-roads lobby has been harried by a ragged band of activists with few of the resources necessary to provide impressive lunches. In March 1997, the *Daily Mail* reported that Steven Norris, Transport Minister for London when John Major's government had been in power, 'made it clear that he shared the views of the "eco-warriors" who took to the tree tops' (quoted in Deans 1997). Norris stated that traffic growth was unsustainable and the construction of the Newbury Bypass had been a mistake. Norris's comments should be viewed in context. He rejected direct action as a means of influencing government policy, stating: 'I agreed with what they said, not what they did. I dislike that kind of eco-fascism which interferes with the democratic process.' (Norris also subsequently worked for a pro-roads group.) Yet the *Mail* article is one sign of the impact that the movement had, indicating its capacity to transform attitudes and challenge policy.

Measuring such apparent success may be difficult and, as we have seen, the impact of the anti-roads movement of the 1990s must be compared with the victories, practical and cultural, of earlier campaigners. Despite Hamer's pessimism, John Tyme in the 1970s directly stopped more roads than have the myriad 1990s campaigners and was the indirect progenitor of the comic novel which brought Blott's ecotage to thousands of readers! Nonetheless the anti-roads movement of the 1990s has advanced green political goals in perhaps a more fundamental manner than did the 1970s movement, by encouraging thousands, rather than Tyme's hundreds, of individuals to practise direct action, as well as by linking transport issues to a deeper social critique. Such success, in turn, would have been impossible without the mobilising ability of EF! (UK) which has acted as a catalyst for direct action and linked protest to radical critique.

The anti-roads movement of the 1990s has generally rejected conventional lobbying and the quest for electoral influence, yet has helped to influence the policy process, as the Norris story indicates. The roads programme proposed in the DoT's *Roads for Prosperity* has been massively reduced, and both former-government ministers like Norris and the new Labour government increasingly emphasise traffic reduction rather than road construction. Economic recession has, of course, been influential, but the recovery in growth since 1993 has seen no renewal of enthusiasm for road-building.

Yet many activists argue that direct action should not be reduced to a mediating stage in the policy process. Steve Booth, one of three editors of *Green Anarchist* convicted and imprisoned in 1997 on a charge of promoting acts of sabotage (albeit in the name of environmental concern and animal rights), noted:

> Most political groups, even including the ALF, aim at negotiation with the government, using their activity as a lever to encourage the state to change its policies. I have repeatedly said...this tactic is futile. The political system does not recognise any interests outside its own. CCTV in every street, computer data bases, phone taps, the whole economic system dedicated to destroying the earth's resources, and an increasingly moronic mass culture aimed at annihilating all individuality; this vast Machine can *never* be negotiated with, or persuaded into some kind of 'Softer Gentler Ecocide'. It can

> only be dismantled, physically destroyed and culturally undermined. Trials and prison do not refute this truth.
>
> (Booth 1997)

Whether conceptualised as 'direct' or 'mediated', the political equation in favour of direct action would clearly be quite different within a more open policy process. As Marshall noted:

> Friends in The Netherlands...go and do a direct action, and the next thing is the Minister says: 'Well, lets all come in and talk about it...'.They can't form a large radical movement because the government immediately absorbs them and starts listening to them and accommodating what they say, and they get results. The Netherlands is a much better and in many ways much greener society than Britain as a result of the fact that government listens. The flip-side of that is they can't form any form of social movement. In Britain we have a government that doesn't listen to anything, but the flip-side of that is that we have the conditions...the basis for Earth First!
>
> (Marshall in interview)

The anti-roads movement has also promoted cultural change in a way that has proved challenging for environmental pressure groups like FoE or Greenpeace, which may fear making demands that alienate potential donors. Living as they do on more subtle resources, both EF! (UK) and the wider anti-roads movement have been able to call for more fundamental transformation, without having to frame their demands so as to appease car owners.

While anti-roads protest has had an impact on the policy process, activists should not be complacent. Measured in terms of transformed political opportunities, success has been limited. The movement is a product of political closure and such closure remains unbreached in the late 1990s. New allies such as trade unionists and young party-goers have been attracted to environmental direct action as much because of their own marginalised positions as by the dynamism of new forms of green mobilisation. Even FoE's Director was sceptical that a Blair government would champion political openness: 'Is he going to repeal the Criminal Justice Act? Is he going to tear down the iron gates outside Number 10, as he promised to do? I doubt it' (Secrett in interview).

While several European Green Parties participate in government or are considered, at the very least, to be serious political players, there are few indications that electoral change will introduce such influence for the Green Party in the UK. Like the Major and Thatcher governments, Blair's administration is productivist in orientation, sympathetic to business, committed to enhanced economic growth and suspicious of pressure groups, let alone more radical networks like EF! (UK). 1997 saw controversial Labour government support for, among other things, formula-one motor racing and genetic engineering. Again, a Secrett remark is apposite: 'One of the parallels they have with Thatcher is [that] they are a party, or the hierarchy [of the party], for and of business.' Better public transport is a means of enhancing growth rather than a stepping-stone to ecotopia. Thus New Labour is prepared to limit car use as a means of promoting more efficient economic development, but shows little enthusiasm for promoting EF! (UK)'s demands or even many of those of FoE. While the roads programme has been halted, a more fundamental transformation of policy or political opportunity has not occurred.

Agents of change

Questions of strategy and success are closely related to the issue of agency formation. In turn, discussion of green *agency* asks how individuals become supporters of green movements and whether a particular sector of society is predisposed to support green demands. Such discussion has often taken a rather abstract form, and an overt interest in green demands, or even an observed adherence to such demands, should not be mistaken for genuine green agency. Green identity may give rise to support for green parties or green movements, but green goals (see Chapter 1) may require an actively green population that participates in local decision-making and grassroots change. Both the difficulties inherent in the transition to a green society and the need to promote forms of self-management as green alternatives to the market and centralised state-control within such a society suggest that green agency may require the creation not only of a 'green' but of an 'activist' identity (Roberts 1979; Wall 1990).

The EF! (UK) activists' interviews considered earlier suggest that individuals become green activists via a process of increasing involvement, which incorporates environmental and political framing, network membership and increasing levels of participation. Such identity formation was closely linked to changes in economic and political opportunity.

Some, but by no means all, of the activists interviewed had been influenced by cultural representations of green political concern such as televised programmes or remembered experience of suddenly imposed environmental grievance during childhood. Many activists came from families in which parents or guardians had acted as social or political entrepreneurs, taking roles variously as church elders, local councillors or workplace shop-stewards. Adult activist identity was cemented by green cultural practices (see Chapter 7). Thus Goodin's suggestion that green-lifestyle commitment is a potential distraction from green political activity ignores the fact that green-agency formation demands a transformation of identity and assumptions that may be promoted by such lifestyle commitment (1992: 83).

The urgency felt by EF! (UK) activists, together with their largely undogmatic approach, their emphasis on personal empowerment and their ability to link 'militant particularist' struggles to global concerns, are important aids to green-agency/identity formation. Environmental pressure groups may be able to frame their demands so as to mobilise financial support or maintain letter-writing campaigns, but seem far less effective in transforming public opinion in a more fundamental way or in promoting the growth of green agency. Environmental pressure groups, irrespective of the extent of their direct influence on the policy process, which have little time for individual amateur entrepreneurs may even impair the formation of green agency by discouraging supporters from the intense involvement and the culture of activism that create green identity. Hannigan (1995: 44) notes:

> Environmental claim-makers are more likely to take the form of professional social movements with paid administrative and research staffs, sophisticated fund-raising programmes and strong, institutionalised links both to legislators and [to] the mass media. Some groups even use door-to-door canvassers who are paid an hourly wage or get to keep a percentage of their solicitations. Campaigns are planned in advance, often in pseudo-military fashion. Grassroots participation is not encouraged beyond 'puppet memberships' with control centralised in the hands of a core group of full-time activists.

Demobilisation and decentred identities

Identities, green or otherwise, are rarely fixed. EF! (UK) and the wider anti-roads movement have mobilised in a particular context, and the effectiveness of green direct action may diminish with changed circumstances. Political change, reduction in youth unemployment, diminished visibility for environmental problems and altered cultural factors could all, separately or in combination, retard future mobilisation. For example, accelerated time–space compression, as well as aiding the diffusion of green ideas and promoting cultural change and hybridisation, may decentre green identities, a process that may be particularly corrosive to ideologically 'loose' networks such as EF! (UK). Indeed, as direct action against road construction has grown, so the influence of EF! (UK) has lessened in importance within the wider movement. Technologies used by environmental activists to increase efficiency during direct actions, such as the mobile phone and the internet, may prove more significant in propelling the environmentally destructive accumulation that greens seek to slow and ultimately reverse.

The economic opportunities that have helped fuel activism may also be closing. Student grants have been replaced by loans, dissuading school-leavers from taking up places in higher education and making the search for employment an absorber of the time which in the past was given over to protest. Youth unemployment may be reduced by economic growth, but so also will individual availability for activist involvement. New Labour's 'New Deal' along with Major's Job Seeker's Allowance are further reducing discretionary time. Activists, despite joining campaigns against the Job Seeker's Allowance, have a poor record of resisting fiscal threats to their biographical availability.

A certainty of social-movement theory is that waves of mobilisation ebb and flow; they do not grow in a continuous and sustained fashion. Yet EF! (UK) has promoted radical repertoires of action and forged a wave of green protest even after a period of decline in environmental activism. New targets for direct action have emerged. Direct action has become more acceptable to both greens and wider society, a move that must please the most radical opponents of planetary destruction. As Reclaim the Streets has argued:

> Direct action enables people to develop a new sense of self-confidence and an awareness of their individual and collective power. Direct action is founded on the idea that people can develop the ability for self-rule only through practice, and proposes that all persons directly decide the important issues facing them. Direct action is not just a tactic, it is individuals asserting their ability to control their own lives and to participate in social life without the need for mediation or control by bureaucrats or professional politicians. Direct action encompasses a whole range of activities, from organising co-ops to engaging in resistance to authority. Direct action places moral commitment above positive law. Direct action is not a last resort when other methods have failed, but the preferred way of doing things.
>
> (RTS leaflet distributed in July 1996)

New networks, new tactics and new styles of protest have been initiated that will doubtless fuel future green activism. From the early green politics of the 1880s to the Kinder Scout Trespass in the 1930s, to Greenham Common in the 1980s, and on to Twyford, the M11 and Newbury, green political protest has changed, grown and will continue to resist new threats, to the environment and to society, borrowing symbolic resources and repertoires from previous episodes of action. As Abbey's Doc Sarvis, who launched the monkeywrench gang's campaign of ecotage, noted: 'Pan shall rise again' (Abbey 1991: 54). Literature is one source of very real revolutionary energy, but narratives change in reaction to transformed natural and social circumstances, and the old god/dess is represented by different forms. While green identities may be in flux, mutation aids reproduction, just as in pagan theology, and, I would suggest, provides a source of renewed hope. EF! (UK) will not be the last Osiris.

Appendix

Contact addresses

Earth First! (UK) can be contacted at:

Earth First! Action Update
Dept 29
1 Newton Street
Manchester M1 1HW
Tel 0161 224 4846

Earth First! (US) can be contacted at:

Earth First! Journal
POB 1415
Eugene
Oregon
OR 97440
USA

Reclaim the Streets can be contacted at:

RTS
PO Box 9656
London N4 4JH

The MOVE organisation can be contacted at:

MOVE
PO Box 19709
Philadelphia
Pennsylvania
PA 19143
USA

Green Anarchist is available from:

BCM 1715
London WC1N 3XX

Do or Die! is available from:

South Downs EF!
c/o Prior House
6 Tilbury Place
Brighton BN2 2GY

Larry O'Hara's publications are available from:

The Green Party Anti-Racist and Anti-Fascist Network
The Green Party
1a Waterloo Road
London W19 5NJ

Bibliography

Abbey, E. (1991) *The Monkey Wrench Gang*, London: Robin Clark.
Abu-Jamal, M. (1995) *Live from Death Row*, New York: Addison-Wesley.
Adorno, T., Brunswick, E., Levinson, D. and Sandford, R. (1950) *The Authoritarian Personality*, New York: Harper.
Amin, A. (ed.) (1995) *Post-Fordism: A Reader*, Oxford: Basil Blackwell.
Anderson, P. (1993) 'Quite the most vicious feud...', *New Statesman and Society*, 13 October: 19.
Andrews, T. (1991) 'Making headway', *Green Line* 87: 4–5.
Anings, N. (1980) *Squatting: the Real Story*, London: Bayleaf.
Anon. (1982) 'M40 fights on', *Undercurrents* 57: 5.
Anon. (1992) 'Road fury', *Wild* 1: 2.
Anon. (1993) 'Dancing on the edge', *Green Revolution* 4: 5–9.
Anon. (1994) 'Auto-struggles: the developing war against the road monster', *Aufheben* 3: 3–23.
Anon. (1996a) 'Earth First! Finland', *Direct Action Direct!* 3: 10–11.
Anon. (1996b) 'Review: *Senseless Acts of Beauty: Cultures of Resistance since the Sixties*', *Aufheben* 5: 43–4.
Armstrong, S. (1996) 'Raging sound of dissent', *Guardian*, 30 September, Section 2: 13.
Atton, C. (1996) 'Anarchy on the internet', *Anarchist Studies* 4: 115–32.
Bagguley, P. (1992) 'Protest, poverty and power: a case study of the anti-Poll Tax movement', *Sociological Review* 43(4): 693–719.
——(1995) 'Social change, the middle class and the emergence of "new social movements": a critical analysis', *Sociological Review* 40(1): 26–48.
Bahro, R. (1984) *From Red to Green*, London: Verso.
Bari, J. (1992) 'The feminization of Earth First!', *Green Revolution* 2: 9.
—— (1993) 'Earth First! in Northern California: interview with Judi Bari', *Capitalism, Nature, Socialism* 4(4): 1–29.
—— (1994) *Timber Wars*, Monroe, ME: Common Courage Press.
Barr, J. (ed.) (1971) *The Environmental Handbook*, London: Friends of the Earth/Ballantine.
Begg, A. (1992) 'Witness for the world', *Green Line* 94: 9–10.
Bellos, A. and Vidal, J. (1996) 'Trees planted in fast lane at protesters' party', *Guardian*, 15 July: 7.
Bennie, L., Franklin, M. and Rüdig, W. (1995) 'Green dimensions', *Green Politics Three*, Edinburgh: Edinburgh University Press.
Benton, E. and Redclift, M. (eds) (1994) 'Introduction', *Social Theory and the Global Environment*, London: Routledge.

Berans, C. (1997) 'Are you fluffy or spiky?', *The Big Issue* 219: 6–7.
Bey, H. (1985) *T.A.Z*, New York: Autonomedia.
Bhaskar, R. (1989) *Reclaiming Reality*, London: Verso.
Bird, T. (1996) 'M11/Claremont Road revisited', unpublished paper delivered at the *New Statesman*'s 'CityStates: Signs of the times' conference, 29 June.
Blaikie, N. (1993) *Approaches to Social Inquiry*, Cambridge: Polity.
Blühdorn, I., Krause, F. and Scharf, T. (eds) *The Green Agenda: Environmental Politics and Policy in Germany*, Keele: Keele University Press.
Bookchin, M. (1988) 'Social ecology versus deep ecology', *Socialist Review* 18(3): 9–29.
Bookchin, M. and Foreman, D. (1991) *Defending the Earth*, Boston, MA: South End Press.
Booth, S. (1996) *Into the 1990s with* Green Anarchist, Oxford: Green Anarchist Books.
—— (1997) 'Statement in response to the Gandalf Trial, Portsmouth', unpublished manuscript.
Borgmann, A. (1995) 'The nature of reality and the reality of nature', in M. Soule and G. Lease (eds) *Reinventing Nature?*, Washington, DC: Island Press.
Bosso, C. (1994) 'Review: *The Politics of Transformation – Local Activism in the Peace and Environmental Movements*', *Journal of Politics* 56(1): 272–4.
Bradford, G. (1989) *How Deep is Deep Ecology?*, Ojai, CA: Times Change Press.
Braungart, R. (1971) 'Family status, socialization and student politics: a multivariate analysis', *American Journal of Sociology* 77: 108–29.
Brookes, S. and Richardson, J. (1975) 'The environmental lobby in Britain', *Parliamentary Affairs* 28: 312–28.
Brookes, S. and Richardson, J. (1976) 'The growth of the environment as a political issue in Britain', *British Journal of Political Science* 6: 244–55.
Brown, P. (1995) 'Twyford M3 protesters win £50,000', *Guardian*, 6 January: 7.
Brown, P. and Mastersonallen, S. (1994) 'The toxic waste movement – a new type of activism', *Society and Natural Resources* 7(3): 269–87.
Bryant, B. (1996) *Twyford Down*, London: Chapman and Hall.
Burbridge, J. (1991) 'The forest is almost gone', *Green Line* 92: 10.
—— (1992a) 'Global action against rainforest destruction', *Green Revolution* 2: 6–7.
—— (1992b) 'Murray Bookchin – eco-guru?', *Wild* 1: 9.
Burklin, W. (1987) 'Governing left parties frustrating the radical non-established left: the rise and inevitable decline of the greens', *European Sociological Review* 3(2): 109–26.
Burns, R. and van der Will, W. (1989) *Protest and Democracy in West Germany*, London: Macmillan.
Byrne, P. (1988) *The Campaign for Nuclear Disarmament*, London: Croom Helm.
Capra, F. (1984) *The Turning Point*, London: Flamingo.
Capuzza, J. (1992) 'A critical analysis of image management within the environmental movement', *The Journal of Environmental Education* 24: 9–14.
Charlesworth, G. (1984) *A History of British Motorways*, London: Thomas Telford.
Charman, P. (1992) 'How world extremists set up havens in London', *Evening Standard*, 15 October: 15.
Cheney, J. (1987) 'Eco-feminism and deep ecology', *Environmental Ethics* 9(2): 115–45.
Clausen, B. and Pomeroy, D. (1994) *Walking on the Edge: How I Infiltrated Earth First!*, Washington: Washington Contract Loggers' Association.
Cloward, R. and Piven, F. (1984) 'Disruption and organization: a rejoinder to Gamson and Schmeidler', *Theory and Society* 13: 587–99.
Cohen, J. (1985) 'Strategy and identity: new theoretical paradigms and contemporary social movements', *Social Research* 52: 663–716.

Cohen, J. and Arato, A. (1992) *Civil Society and Political Theory*, Cambridge, MA: MIT Press.
Cohen, N. (1992) 'Eco-radicals warn of violence', *Independent on Sunday*, 19 April: 5.
Collin, M. and Godfrey, J. (1997) *Altered State*, London: Serpent's Tail.
Control, S. (1991) *Away with All Cars*, Stoke-on-Trent: Play Time for Ever Press.
Cooper, S. (1992) 'The ideology and practice of Earth First!', unpublished manuscript, University of Bristol.
Copjec, J., (1995) *Read My Desire*, London: MIT.
Cosgrove, S. (1988) 'Forbidden fruits', *New Statesman and Society* 1(13): 44.
Costello, A. (1992) 'The ecology of failure', *Analysis* 1(1): 2–9.
Cotgrove, S. (1982) *Catastrophe or Cornucopia*, London: John Wiley.
Cotgrove, S. and Duff, A. (1980) 'Environmentalism, middle-class radicalism and politics', *Sociological Review* 28: 333–51.
Cracknell, J. (1993) 'Issue arena, pressure groups and environmental agendas', in F. Cylke (ed.) *The Environment*, New York: HarperCollins College Publishers.
Crossley, N. (1995) 'Watching us, watching you', *The Big Issue* 127: 14.
Cylke, F. (ed.) (1993) *The Environment*, New York: HarperCollins College Publishers.
Daniels, S. (1994) *Fields of Vision*, Oxford: Polity.
Davis, J. (1971) *When Men Revolt and Why*, New York: Free Press.
—— (ed.) (1991) *The Earth First! Reader*, Salt Lake City, UT: Gibbs Smith.
Day, D. (1989) *The Eco Wars*, London: Harrap.
De Nardo, J. (1985) *Power in Numbers*, Princeton, NJ: Princeton University Press.
Deans, J. (1997) ' "Road protest was right" says Norris', *Daily Mail*, March 17: 13.
Della Porta, D. (1992) 'Life histories in the analysis of social movement activists', in M. Diani and R. Eyerman (eds) *Studying Collective Action*, London: Sage.
—— (1995) *Social Movements, Political Violence and the State*, Cambridge: Cambridge University Press.
Devall, W. (1985) *Deep Ecology: Living as if Nature Mattered*, Salt Lake City, UT: Peregrine Smith.
Diani, M. (1994) *Green Networks*, Edinburgh: Edinburgh University Press.
Diani, M. and Eyerman, R. (1992) *Studying Collective Action*, London: Sage.
Dickens, P. (1992) *Society and Nature – Towards a Green Social Theory*, Hemel Hempstead: Harvester Wheatsheaf.
Dickson, B. (1992) 'A crisis of diversity', *Green Line* 101: 10–11.
Dickson, D. (1974) *Alternative Technology and the Politics of Technical Change*, London: Fontana.
Dobson, A. (1990) *Green Political Thought*, London: Unwin Hyman.
Doherty, B. (1992) 'The Fundi-realo controversy', *Environmental Politics* 1(1): 95–120.
Doherty, B. and Rawcliffe, P. (1995) 'British exceptionalism? Comparing the environmental movement in Britian and Germany', in I. Blühdorn, F. Krause, and T. Scharf (eds) *The Green Agenda: Environmental Politics and Policy in Germany*, Keele: Keele University Press.
Downes, L. (1992) 'Green revolt of the ordinary citizens', *Independent on Sunday*, 5 July: 10.
Downs, A. (1972) 'Up and down with ecology – the issue attention cycle', *Public Interest* 28: 38–50.
Dudley, G. and Richardson, J. (1996) *Arenas as Agents of Policy Change: New Institutionalism and the Public Inquiry Process in UK Trunk Roads Policy*, Essex Papers in Politics and Government, no.104, Colchester: Essex University.
Duncan, A. (1992) *Taking on the Motorway*, London: Kensington and Chelsea Community History Group.

Dunlap, R. and Mertig, A. (1991) 'The evolution of the United States' environmental movement from 1970 to 1990 – an overview', *Society and Natural Resources* 4 (3): 209–18.

Dunlap, R. and Scarce, R. (1990) 'The polls: poll trends, environmental problems and protection', *Public Opinion Quarterly* 55: 651–6.

Ecologist, The (1972) *A Blueprint for Survival*, Harmondsworth: Penguin.

—— (1992) 'Whose common future?', *The Ecologist* 22: 4.

Economist (1994) 'The classless society', February 19: 27–8.

Edwards, B. (1995) 'With liberty and environmental justice for all', in B. Taylor (ed.) *Ecological Resistance Movements*, New York: State University of New York Press.

Ehrenfeld, D. (1978) *The Arrogance of Humanism*, New York: Oxford University Press.

Ehrlich, P. (1968) *The Population Bomb*, New York: Ballantine.

Eisinger, P. (1973) 'The conditions of protest behaviour in American cities', *American Political Science Review* 67: 11–28.

Elkins, S. (1989) 'The politics of mystical ecology', *Telos* 82: 52–70.

Ellis, R. and Coyle, D. (1994) *Politics, Policy and Culture*, Boulder, CO: Westview.

Emirbayer, M. and Goodwin, J. (1994) 'Network analysis, culture, and the problem of agency', *American Journal of Sociology* 99: 1411–54.

Evans, D. (1992) *A History of Nature Conservation*, London: Routledge.

Evans, G. (1993) 'Hard times for the British Green Party', *Environmental Politics* 2(2): 327–33.

Evans, J. (1975) *The Environment of Early Man in the British Isles*, London: Paul Elek.

Eyerman, R. and Jamison, A. (1989) 'Environmental knowledge as an organizational weapon: the case of Greenpeace', *Social Science Information* 28(1): 99–119.

—— (1991) *Social Movements: A Cognitive Approach*, Cambridge: Polity.

Fallon, I. (1991) *Billionaire*, London: Hutchinson.

Fantasia, R. (1988) *Cultures of Solidarity*, Berkeley, CA: University of California Press.

Fermia, J. (1994) 'Political culture', in W. Outhwaite, and T. Bottomore (eds) *The Blackwell Dictionary of Twentieth-Century Social Thought*, Oxford: Basil Blackwell.

Fernadez, R. and McAdam, D. (1987) 'Multiorganizational fields and recruitment to social movements', in B. Klandermans (ed.) *Organizing for Social Change: Social Movement Organizations Across Cultures*, Greenwich, CT: JAI Press.

Field, P. and Drury, J. (1995) *Claremont Road E11 – A Festival of Resistance*, Leyton, London: No M11 Link.

Fielding, N. (1993) 'Qualitative interviewing', in N. Gilbert (ed.) *Researching Social Life*, London: Sage.

Fielding, N. and Fielding, J. (1986) *Linking Data – The Articulation of Qualitative and Quantitative Methods in Social Research*, London: Sage.

Finch, R. (1991) 'The case against the car', *Green Line* 90: 9–10.

Finch, S. and Peltz, L. (1992) *21 Years of Friends of the Earth*, London: Friends of the Earth.

Flynn, A. and Lowe, P. (1992) 'The greening of the Tories', *Green Politics Two*, Edinburgh: Edinburgh University Press.

Foreman, D. (1991) *Confessions of an Eco-Warrior*, New York: Harmony.

Foreman, D. and Haywood, B. (1993) *Ecodefense*, Chico, CA: Abzug Press.

Foreman, D. and Morton, N. (1991) 'Good luck, darlin', it's been great', in J. Davis (ed.) *The Earth First! Reader*, Salt Lake City, UT: Gibbs Smith.

Forster, E.M. (1965) *Two Cheers for Democracy*, Harmondsworth: Penguin.

Foweraker, J. (1995) *Theorising Social Movements*, London: Pluto.

Fox, W. (1990) *Towards a Transpersonal Ecology*, Boston: Shambala.

Frankel, B. (1987) *The Post-Industrial Utopians*, Cambridge: Polity.
Frankland, E. (1990) 'Does green politics have a future in Britain?', *Green Politics One*, Edinburgh: Edinburgh University Press.
Freeman, J. (1972–3) 'The tyranny of structurelessness', *Berkeley Journal of Sociology* 17: 151–64.
Freeman, J. (1980) *Social Movements of the Sixties and Seventies*, New York: Longman.
Friedman, D. and McAdam, D. (1992) 'Collective identity and activism: networks, choices, and the life of a social movement', in A. Morris and C. Mueller (eds) *Frontiers in New Social Movement Theory*, New Haven, CT: Yale University Press.
Friends of MOVE (1996) *25 Years on the Move*, London: Friends of MOVE.
Freudenberg, N. (1984) *Not in Our Backyards*, New York: Monthly Review Press.
Gale, R. (1986) 'Social movements and the state – the environmental movement, countermovement, and government agencies', *Sociological Perspectives* 29(2): 202–40.
Gallagher, J. (1995) 'Unguarded behaviour', *New Statesman and Society*, 31 March: 22–4.
Gamson, W. (1975) *The Strategy of Social Protest*, Homewood, IL: Dorsey.
—— (1992) 'The social psychology of collective action', in A. Morris and C. Mueller (eds) *Frontiers in New Social Movement Theory*, New Haven, CT: Yale University Press.
Gandalf (1993) 'The realization and suppression of the Earth Liberation Front', *Green Revolution* 4: 4–5.
Garner, R. and Zald, M. (1985) 'The political economy of social movement sectors', in G. Suttles and M. Zald (eds) *The Challenge of Social Control*, Norwood, NJ: Ablex.
Gartment, D. (1994) *Auto Opium*, London: Routledge.
Gedicks, A. (1995) 'International native resistance to the new resource wars', in B. Taylor (ed.) *Ecological Resistance Movements*, New York: State University of New York Press.
Gerlach, L. and Hine, V. (1970) *People, Power and Change*, Indianapolis, IN: Bobbs-Merrill.
Giddens, A. (1987) *Social Theory and Modern Sociology*, London: Heinemann.
Goodin, R. (1992) *Green Political Theory*, Cambridge: Polity.
Goodman, A. (1995) 'The Criminal Justice and Public Order Act 1994', *Capital and Class* 56: 9–13.
Goodwin, N. (1995) 'Jackboot and the beanfield', *New Statesman*, 23 June: 22–3.
Gould, K., Weinberg, A. and Schnaiberg, A. (1993) 'Legitimating impatience: pyhrric victories of the modern environmental movement', *Qualitative Sociology* 16(3): 207–46.
Gould, P. (1988) *Early Green Politics*, Brighton: Harvester.
Granovetter, M. (1973) 'The strength of weak ties. A network theory revisited', *American Journal of Sociology* 78: 1360–80.
Grant, J. (1977) *The Politics of Urban Transport Planning*, London: Earth Resources Research.
Green 2000 (1990) 'Towards a green 2000', unpublished manuscript, St Margaret's, Middlesex.
Greenpeace International (1991) *The Environmental Impact of the Car*, London: Greenpeace.
Gregory, R. (1974) 'The Minister's line: or, the M4 comes to Berkshire', in R. Kimber and J. Richardson (eds) *Campaigning for the Environment*, London: Routledge and Kegan Paul.
Gurney, J. and Tierney, K. (1982) 'Relative deprivation and social movements: a critical look at twenty years of theory and research', *Sociological Quarterly* 23: 33–47.
Gurr, E. (1970) *Why Men Rebel*, Princeton, NJ: Princeton University Press.
Hall, I. (1976) 'Community action versus pollution', Social Science Monograph no.2, University of Wales, Cardiff.

Hall, P. (1980) *Great Planning Disasters*, London: Weidenfeld & Nicolson.
Hall, S. (1985) 'Religious ideologies and social movements in Jamaica', in R. Bocock and K. Thompson, (eds) *Religion and Ideology*, Manchester: Manchester University Press.
Hall, S. (1994) 'A criminally unjust bill', *Red Pepper* 6: 36–7.
Hall, S., Clarke, J., Jefferson, T. and Roberts, B. (eds) (1976) *Resistance Through Rituals*, London: Hutchinson.
Hamer, M. (1987) *Wheels Within Wheels*, Andover, Hants: Routledge & Kegan Paul.
Hannigan, J. (1995) *Environmental Sociology*, London: Routledge.
Harries-Jones, P. (1993) 'Between science and shamanism', in K. Milton (ed.) *Environmentalism: The View from Anthropology*, London: Routledge.
Harris, D. (1992) *From Class Struggle to the Politics of Pleasure*, London: Routledge.
Harry, M. (1987) 'Attention MOVE! This is America', *Race and Class* 28(4): 5–28.
Hart, D. (1976) *Strategic Planning in London*, Oxford: Pergamon.
Harvey, D. (1990) *The Condition of Postmodernity*, Oxford: Basil Blackwell.
—— (1996) *Justice, Nature and the Geography of Difference*, Oxford: Basil Blackwell.
Hays, S. (1987) *Beauty, Health and Permanence: Environmental Politics in the United States: 1955–1985*, London: Cambridge University Press.
Hebdige, D. (1979) *Subculture*, London: Methuen.
Heffernan, R. and Marquesse, M. (1992) *Defeat from the Jaws of Victory*, London: Verso.
Hendriks, F. (1994) 'Cars and culture in Munich and Birmingham', in R. Ellis and D. Coyle, (eds) *Politics, Policy and Culture*, Boulder, CO: Westview.
Hepple, T. (1993) *At War with Society: The Exclusive Story of a Searchlight Mole Inside Britain's Far Right*, London: Searchlight.
Hill, B., Freeman, R., Blanire, S. and McIntosh, A. (1995) 'Popular resistance and the emergence of radical environmentalism in Scotland', in B. Taylor (ed.) *Ecological Resistance Movements*, New York: State University of New York Press.
Hirst, P. (1989) *After Thatcher*, London: Collins.
Hoffer, E. (1951) *The True Believer*, New York: Harper & Row.
Hogg, S. and Hill, T. (1995) *Too Close to Call*, Boston: Little, Brown & Co.
Holdsworth, D. (1975) *The New Society*, Reading: Thames Valley Police.
Honigsbaum, M. (1994) Force of nature, *GQ*, May, 111–13 and 154.
—— (1996) 'Why I want to block the capital's roads', *Evening Standard*, 16 July: 28.
Hook, W. (1993) 'A view from the hilltop', *Green Line* 104: 10–11.
Howe, D. Stammers, M. and Walker, J. (1994) *Dr Who – the Seventies*, London: Virgin.
Huberman, A. and Miles, M. (1985) 'Assessing local causality in qualitative research', in D. Berg and K. Smith (eds) *Exploring Clinical Methods for Social Research*, Beverly Hills, CA: Sage.
Huey, S. (1990) *Annual Report and Accounts, 1989–1990*, London: Friends of the Earth.
Hunt, H. (1995) 'Balancing act: personal politics and anti-roads campaigning', *Soundings* 1(1): 123–38.
Huq, R. (1995) 'The culture of rebellion', *Red Pepper* 3: 14–15.
Inanna [David Taylor] (1983) 'The greens are gathering', *Peace News* 2195: 15–16.
Inglehart, R. (1977) *The Silent Revolution*, Princeton, NJ: Princeton University Press.
Irvine, S. and Ponton, A. (1988) *A Green Manifesto*, London: Optima.
Isaac, J. (1987) *Power and Marxist Theory*, Ithaca, NY: Cornell University Press.
Jackson, M., Peterson, E., Bull, J., Monson, S. and Richmond, P. (1960) 'The failure of an incipient social movement', *Pacific Sociological Review* 31: 35–40.
Jasper, J. and Poulsen, J. (1995) 'Recruiting strangers and friends: moral shocks and social networks in animal rights and anti-nuclear protests', *Social Problems* 42(4): 493–512.

Jenkins, J. (1983) 'Resource mobilization theory and the study of social movements', *Annual Review of Sociology* 82: 527–53.

Jenkins, J. and Perrow, C. (1977) 'Insurgency of the powerless: farm worker movements (1945–1972)', *American Sociological Review* 42(2): 249–68.

Jimmie, K. and Palmer, J. (1992) *Ecospeak*, Southern Illinois: Southern Illinois University Press.

Johnson, S. (1974) *The Politics of Ecology*, London: Tom Stacey.

—— (1981) *Caring for the Environment*, London: Conservative Political Centre and European Democratic Group.

Jones, C. (1986) 'Shelter from the storm: under the green umbrella', in G. Chester and A. Rigby (eds) *Articles of Peace*, Bridport, Dorset: Prism Press.

Jones, D. (1979) 'Not in my community', *Social Policy* 10(2): 44–6.

Jones, L. (1983) *Keeping the Peace*, London: Women's Press.

Jordan, J. (1996) 'M11/Claremont Road Revisited', unpublished paper delivered at the *New Statesman*'s 'CityStates: Signs of the times' conference, 29 June.

Jordan, T. (1994) *Reinventing Revolution*, Aldershot: Avebury.

Jowers, P. (1994) 'Towards the politics of a "lesser evil": Jean-Francois Lyotard's reworking of the Kantian sublime', in J. Weeks (ed.) *The Lesser Evil and the Greater Good*, London: Rivers Orum Press.

Kemp, P. and Wall, D. (1990) *A Green Manifesto for the 1990s*, Harmondsworth: Penguin.

Kendell, J. and Buivids, E. (1987) *Earth First*, Sydney: Associated Broadcasting Corporation.

Keniston, K. (1968) *Young Radicals*, New York: Harcourt, Brace & World.

Kimber, R. and Richardson, J. (1974) *Campaigning for the Environment*, London: Routledge & Kegan Paul.

Kine, B. (1997) *First Along the River*, San Francisco, CA: Acada Books.

Kingdon, J. (1994) *Agenda, Alternatives and Public Policies*, Boston: Little, Brown & Co.

Kingston, B. (1996) *Laura's Way*, London: Arrow.

Kitschelt, H. (1985) 'New social movements in West Germany and the United States', in M. Zeitlin (ed.) *Political Power and Social Theory*, Greenwich, CT: JAI Press.

—— (1986) 'Political opportunity structures and political protest: anti-nuclear movements in four democracies', *British Journal of Political Science* 16: 57–85.

—— (1989) *The Logic of Party Formation: Ecological Politics in Belgium and West Germany*, Ithaca, NY: Cornell University Press.

Klandermans, B. (1984) 'Mobilization and participation: social psychological expansions of resource mobilization theory', *American Sociological Review* 49: 583–600.

—— (1990) 'Linking the "old" and the "new": movement networks in The Netherlands', in R. Dalton and M. Kuchler (eds) *Challenging the Political Order*, Cambridge: Polity Press.

Klandermans, B. and Oegema, D. (1987) 'Potentials, networks, motivations, and barriers: steps towards participation in social movements, *American Sociological Review* 52: 503–31.

Klandermans, B. and Tarrow, S. (1988) 'Mobilization into social movements: synthesizing European and American Approaches', *American Sociological Review* 52: 519–31.

Kramer, J. (1988) 'Letter from Europe', *The New Yorker*, 28 November: 67–110.

Kriesi, H. (1992) 'The rebellion of the research "objects" ', in M. Diani and R. Eyerman, *Studying Collective Action*, London: Sage.

Kriesi, H., Koopmans, R., Duyvendak, J. and Giugni, M. (1995) *New Social Movements in Western Europe*, London: UCL Press.

Laclau, E. and Mouffe, C. (1985) *Hegemony and Socialist Strategy: Towards a Radical Democratic Politics*, London: Verso.

Lamb, R. (1996) *Promising the Earth*, London: Routledge.

Lambert, J. (1991) 'Letter', *Green Line* 88: 23.

Lang, C. (1992) 'Non au tunnel du Somport!', *Green Line* 98: 19.

Lang, T. and Raven, H. (1994) 'From market to hypermarket', *The Ecologist* 24(4): 124–37.

Larana, E., Johnston, H. and Gusfield, J. (1994) *New Social Movements: From Ideology to Identity*, Philadelphia, PA: Temple University Press.

Lash, S. and Urry, J. (1987) *The End of Organised Capitalism*, Oxford: Polity.

Lee, M. (1995) *Earth First! Environmental Apocalypse*, New York: Syracuse University Press.

Lee, R. (1983) 'Free the 4,344,843', *Peace News* 2201: 5–6.

—— (1990) 'Animal liberation', *Arkangel* 2: 39.

Lees, A. and Bourn, D. (1992) 'Earth First!', unpublished manuscript, London: Friends of the Earth.

Leopold, A. (1949) *A Sand County Almanac*, New York: Oxford University Press.

Levin, P. (1979) 'Public inquiries: the need for natural justice', *New Society*, 15 November.

Lichterman, P. (1995) 'Piecing together multicultural community', *Social Problems* 42(4): 513–34.

Liddington, J. (1989) *The Long Road to Greenham*, London: Virago.

Lofland, J. (1993) *Polite Protest*, Syracuse, NY: Syracuse University Press.

—— (1994) 'Charting degrees of movement culture: tasks of the cultural cartographer', in H. Johnston and B. Klandermans (eds) *Social Movements and Culture*, London: UCL Press.

Lofland, J., Colwell, M. and Johnson, V. (1990) 'Change-theories and movement structure', in J. Lofland (ed.) *Peace Action in the 80s*, London: Rutgers University Press.

Lowe, P. and Flynn, A. (1989) 'Environmental politics and policy in the 1980s', in J. Mohan (ed.) *The Political Geography of Contemporary Britain*, London: Macmillan.

Lowe, P. and Goyder, J. (1983) *Environmental Groups in Politics*, London: Allen & Unwin.

Lowe, R. and Shaw, W. (1993) *Travellers – Voices of the New Age Nomads*, London: Fourth Estate.

Lukes, S. (1976) *Power – A Radical View*, London: Macmillan.

McAdam, D. (1982) *Political Process and the Development of Black Insurgency, 1930–1970*, Chicago, IL: University of Chicago Press.

—— (1983) 'Tactical innovation and the pace of insurgency', *American Sociological Review* 48: 735–54.

—— (1986) 'Recruitment to high-risk activism: the case of Freedom Summer', *American Journal of Sociology* 92: 64–90.

—— (1988) *Freedom Summer*, New York: Oxford University Press.

—— (1995) ' "Initiator" and "Spin-off" movements: diffusion processes in protest cycles', in M. Traugott (ed.) *Repertoires and Cycles of Collective Action*, London: Duke University Press.

—— (1996) 'Culture and Social Movements', in E. Larana, H. Johnston and J. Gusfield (eds) *New Social Movements: From Ideology to Identity*, Philadelphia, PA: Temple University Press.

McAdam, D. and Fernandez, R. (1990) 'Microstructural bases of recruitment to social movements', *Research in Social Movements, Conflict and Change* 12: 1–33.

McAdam, D., McCarthy, J. and Zald, M. (1988) 'Social movements', in N. Smelser (ed.) *Handbook of Sociology*, Beverly Hills, CA: Sage.

—— (eds) (1996) *Comparative Perspectives on Social Movements*, Cambridge: Cambridge University Press.

McAdam, D. and Paulsen, R. (1993) 'Specifying the relationship between social ties and activism', *American Journal of Sociology* 99: 640–67.

McAdam, D. and Rucht, D. (1993) 'The cross-national diffusion of movement ideas', *The Annals of the American Academy of Political and Social Science* 528: 56–74.

McCarthy, J. and Wolfson, M. (1991) 'The institutional channelling of social movements in the modern state', *Research in Social Movements: Conflict and Change* 13: 45–76.

McCarthy, J. and Zald, M. (1973) *The Trend of Social Movements in America*, Morristown, NJ: General Learning Press.

—— (1977) 'Resource mobilization and social movements: a partial theory', *American Journal of Sociology* 82: 1212–41.

McCormick, J. (1991) *British Politics and the Environment*, London: Earthscan.

McIntosh, M. (1971) 'Changes in the organization of thieving', in S. Cohen (ed.) *Images of Deviance*, Harmondsworth: Penguin.

McKay, G. (1996) *Senseless Acts of Beauty: Cultures of Resistance since the Sixties*, London: Verso.

McManus, M. (1995) *From Fate to Choice: Private Bobbies, Public Beats*, Aldershot: Avebury.

Manes, C. (1990) *Green Rage*, Boston: Little, Brown & Co.

Mann, N. (1986) 'Revolution on the road', *Peace News* 2275: 12–13.

Markovits, A. and Gorski, P. (1993) *The German Left*, Oxford: Polity.

Marsh, J. (1982) *Back to the Land: The Pastoral Impulse in Victorian England from 1880 to 1914*, London: Quartet Books.

Martell, L. (1994) *Ecology and Society*, London: Polity.

Marwell, G., Oliver, P. and Prahl, R. (1988) 'Social networks and collective action', *American Journal of Sociology* 94: 502–34.

Marx, G. (1974) 'Thoughts on a neglected category of social movement participant: the agent provacateur and the informant', *American Journal of Sociology* 80(2): 402–42.

Marx, G. (1979) 'External efforts to damage or facilitate social movements: some patterns, explanations, outcomes, and complications', in M. Zald and J. McCarthy (eds) *The Dynamics of Social Movements*, Cambridge, MA: Winthrop.

Marx, K. (1991) *Capital*, Harmondsworth: Penguin, vol. 3.

Mathew, D. (1987) *Capital Scheme*, London: Friends of the Earth.

Mattausch, J. (1989) *A Commitment to Campaign*, Manchester: Manchester University Press.

May, T. (1994) *The Political Philosophy of Poststructuralist Anarchism*. University Park, PA: Pennsylvania University Press.

Maycock, B. (1987) 'Animal liberationists are peace campaigners, too!', *Peace News* 2289: 7.

Megill, A. (1985) *Prophets of Extremity*, Berkeley: University of California Press.

Melucci, A. (1989) *Nomads of the Present*, London: Radius.

—— (1992) 'Frontier land: collective action between actors and systems', in M. Diani and R. Eyerman (eds) *Studying Collective Action*, London: Sage.

—— (1996) *Challenging Codes*, Cambridge: Cambridge University Press.

Mercer, P. (1994) *The UK Directory of Political Organisations*, London: Longman.

—— (1995) *The UK Directory of Political Organisations*, London: Longman.

Merrick (1996) *Battle for the Trees*, Leeds: Godhaven.

Meyer, D. and Whittier, N. (1994) 'Social movement spillover', *Social Problems* 41(2): 277–98.

Miles, M. and Huberman, A. (1994) *Qualitative Data Analysis*, London: Sage.

Millword, R. and Robinson, A. (1970) *The Lake District*, London: Eyre & Spottiswode.

Morris, A. (1981) 'The Black Southern Sit-in Movement: an analysis of internal organization', *American Sociological Review* 46: 744–67.

Morris, A. and Herring, C. (1987) 'Theory and research in social movement: a critical review', *Annual Review of Political Science* 2: 137–98.

Morris, A. and Mueller, C. (1992) *Frontiers in New Social Movement Theory*, New Haven, CT: Yale University Press.

Mowrey, M. and Redmond, T. (1993) *Not in Our Backyard*, New York: Morrow.

Moyes, J. (1997) 'It's not violence that middle England won't tolerate, it's police snooping', *Independent*, 13 January: 3.

Mueller, C. (1992) 'Building social movement theory', in A. Morris and C. Mueller *Frontiers in New Social Movement Theory*, New Haven, CT : Yale University Press.

Mueller, E. (1980) 'The psychology of political protest violence', in T. Gurr (ed.) *The Handbook of Political Conflict*, New York: Free Press.

Mumford, L. (1970) *The Pentagon of Power*, London: Secker & Warburg.

Naess, A. (1973) 'The shallow and the deep, long-range ecology movement: A summary', *Inquiry* 16: 95–100.

—— (1984) 'A defense of the deep ecology movement', *Environmental Ethics* 6(3): 265–70.

—— (1989) *Ecology, Community and Lifestyle*, London: Cambridge University Press.

—— (1990) 'Man apart and deep ecology – a reply', *Environmental Ethics* 12(2):185–92.

Nash, F. (1989) *The Rights of Nature*, Madison, WI: University of Wisconsin Press.

Nash, R. (1982) *Wilderness and the American Mind*, New Haven, CT: Yale University Press.

Nelkin, D. and Pollak, M. (1982) *The Atom Besieged*, Cambridge, MA: MIT.

Nietzsche, F. (1994) *On the Genealogy of Morality*, London: Cambridge University Press.

Norris, C. (1992) *Uncritical Theory: Postmodernism, Politics, Intellectuals and the Gulf War*, London: Lawrence & Wishart.

Nyhagen Predelli, L. (1995) 'Ideological conflict in the radical environmental group Earth First!', *Environmental Politics* 4(1): 123–9.

Oberschall, A. (1973) *Social Conflict and Social Movements*, Englewood Cliffs, NJ: Prentice-Hall.

Oegema, D. and Klandermans, B. (1994) 'Why social movement sympathizers don't participate: erosion and nonconversion of support', *American Sociological Review* 59: 703–22.

Offe, C. (1985) 'New social movements: challenging the boundaries of institutional politics', *Social Research* 52: 817–68.

O'Hara, L. (1993a) *A Lie Too Far: Searchlight, Hepple and the Left*, London: Mina.

—— (1993b) *At War with the Truth*, London: Mina.

—— (1994) *Turning Up the Heat: MI5 after the Cold War*, London: Phoenix Press.

O'Hara, L. and Matthews, G. (1992) *Paradise Referred Back*, London: Green Flame.

Oliver, P. (1984) ' "If you don't do it, nobody else will": active and token contributors to local collective action', *American Sociological Review* 49: 601–10.

Olson, M. (1965) *The Logic of Collective Action*, Cambridge, MA: Harvard University Press.

Open University (1978) *The Windscale Controversy*, Milton Keynes: Open University Press.

Outhwaite, W. (1987) *New Philosophies of Social Science*, London: Macmillan.

Papadakis, E. (1989) 'Interventions in the new social movements', in J. Gubrium and D. Silverman (eds) *The Politics of Field Research*, London: Sage.

—— (1993) *Politics and the Environment: The Australian Experience*, London: Allen & Unwin.

Parkin, S. (1989) *Green Parties*, London: Heretic.

Paasonen, K.-E. (1983) 'The struggle for the Tasmanian wilderness', *Peace News* 2234: 13.

Patterson, W. (1984) 'A decade of friendship', in D. Wilson (ed.) *The Environmental Crisis*, London: Heinemann.
Paul, L. (1951) *Angry Young Man*, London: Faber.
Payne, D. (1996) *Voices in the Wilderness*, Hanover, New Haven: University Press of New England.
Peace News Collective (1984) 'Non violence and the ALF', *Peace News* 2190: 7.
Pearce, F. (1991) *Green Warriors*, London: Bodley Head.
Pearson, G. (1983) *Hooligan: A History of Respectable Fears*, London: Macmillan.
Pepper, D. (1993) *Eco-Socialism*, London: Routledge.
—— (1996) *Modern Environmentalism*, London: Routledge.
Pilger, J. (1996a) 'They never walk alone', *Guardian*, 23 November: 14–23.
—— (1996b) 'An inspiration to us all', *Red Pepper* 28: 5.
Piven, F. and Cloward, R. (1977) *Poor People's Movements: Why they Succeed and How they Fail*, New York: Pantheon.
Piven, F. and Cloward, R. (1992) 'Normalizing collective protest', in A. Morris and C. Mueller (eds) *Frontiers in New Social Movement Theory*, New Haven, CT: Yale University Press.
Plummer, K. (1983) *Documents of Life*, London: Unwin Hyman.
Ponton, S. and Irvine, S. (1992) 'The crisis in the English Green Party', unpublished paper, Newcastle-Upon-Tyne.
Porritt, J. (1984) *Seeing Green*, Oxford: Basil Blackwell.
Porritt, J. and Winner, D. (1988) *The Coming of the Greens*, London: Fontana.
Pye-Smith, C. and Rose, C. (1984) *Crisis and Conservation*, London: Penguin.
Quodling, P. (1991) *Bougainville: The Mine and the People*, Australia: Centre for Independent Studies.
Rainbow, S. (1993) *Green Politics*, Auckland, New Zealand: Oxford University Press.
Redhead, S. (1993) *Rave Off*, Aldershot: Avebury.
Reich, C. (1973) *The Greening of America*, New York: Bantam.
Richardson, R. (1995) 'The green challenge', in R. Richardson and C. Rootes (eds) *The Green Challenge*, London: Routledge.
Ritetveld, H. (1993) 'Living the Dream', in S. Redhead (ed.) *Rave Off*, Aldershot: Avebury.
Rivers, P. (1974) *Politics by Pressure*, London: Routledge & Kegan Paul.
Roberts, A. (1979) *The Self-Managing Environment*, London: Allison & Busby.
Roberts, J. (1986) *Against All Odds*, London: Arc Print.
Robinson, M. (1992) *The Greening of British Party Politics*, Manchester: Manchester University Press.
Roddewig, R. (1978) *Green Bans*, Sydney: Conservation Foundation.
Rogers, E. (1983) *Diffusion of Innovation*, Detroit, MI: Free Press
Rogers, P. (1992) *Green Anarchism: Its Origins and Influences*, Oxford: Green Anarchist.
Rootes, C. (1981) 'On the future of protest politics in Western democracies – a critique of Barnes, Kase *et al.*, *Political Action*', *European Journal of Political Research* 9: 421–39.
—— (1992) 'The new politics and the new social movements: accounting for British exceptionalism', *European Journal of Political Research* 22: 171–91.
—— (1995) 'Britain: greens in a cold climate', in D. Richardson and C. Rootes (eds) *The Green Challenge*, London: Routledge.
—— (1997) 'Shaping collective action: structure, contingency and knowledge', unpublished manuscripts, University of Kent at Canterbury.
Rose, C. (1993) 'Beyond the struggle for proof: factors changing the environmental movement', *Environmental Values* 2: 285–98.

Roseneil, S. (1993) 'Greenham revisted: researching myself and my sisters', in D. Hobbs and T. May (eds) *Interpreting the Field*, Oxford: Clarendon, 177–208.

—— (1995) *Disarming Patriarchy: Feminism and Political Action at Greenham*, Milton Keynes: Open University Press.

Roszak, T. (1972) *Where the Wasteland Ends*, London: Faber.

Rothman, B. (1982) *The 1932 Kinder Trespass*, Altrincham: Willow.

Rothschild-Whitt, J. (1979) 'The collectivist organization: an alternative to rational-bureaucratic models', *American Sociological Review* 44: 509–27.

Rowell, A. (1996) *Green Backlash*, London: Routledge.

Rucht, D. (1990a) 'Campaigns, skirmishes and battles: anti-nuclear movements in the USA, France and West Germany', *Industrial Crisis Quarterly* 4: 193–222.

—— (1990b) 'The strategies and action repertoires of new movements', in R. Dalton and M. Kuchler (eds) *Challenging the Political Order*, Cambridge: Polity Press.

—— (1995) 'Ecological protest as calculated law-breaking', *Green Politics Three*, Edinburgh: Edinburgh University Press,

Rucht, D. and Ohlemacher, T. (1992) 'Protest event data: collection, uses and perspectives', in M. Diani and R. Eyerman (eds) *Studying Collective Action*, London: Sage.

Rüdig, W. (1990) *Anti-Nuclear Movements*, London: Longman.

—— (1993) 'Editorial', *Green Politics Two*, Edinburgh: Edinburgh University Press, 1–8.

Rüdig, W. and Lowe, P. (1986) 'The withered "greening" of British politics', *Political Studies* 34: 262–84.

Rüdig, W., Bennie, L. and Franklin, M. (1991) *Green Party Members: A Profile*, Glasgow: Delta Publications.

—— (1993) *Green Blues: The Rise and Decline of the British Green Party*, Strathclyde Papers in Government and Politics, no. 94, Department of Government, University of Strathclyde, Glasgow.

Ryder, R. (1989) *Animal Revolution*, Oxford: Basil Blackwell.

Sale, K. (1986) 'The forest for the trees: can today's environmentalist tell the difference?', *Mother Jones* 11: 25–58.

—— (1988) 'The cutting edge – deep ecology and its critics', *Nation* 246(19): 670.

—— (1993) 'Silent Spring and after – the United States' green movement today', *Nation* 257(3): 92–6.

Salleh, A. (1984) 'Deeper than deep ecology – the eco-feminist connection', *Environmental Ethics* 6(4): 339–45.

Salt, H. (1892) *Animals' Rights Connected in Relation to Social Progress*, London: Bell.

Sayer, A. (1986) 'New developments in manufacturing: the just-in-time system', *Capital and Class* 30: 43–72.

Sayer, D. (1984) *Method in Social Science*, London: Hutchinson.

Scarce, R. (1990) *Eco-Warriors*, Chicago, IL: Noble.

Schmitt, R. (1989) 'Organizational interlocks between new social movement and traditional elites: the case of the West German peace movement', *European Journal of Political Research* 17: 583–98.

Schmitt-Beck, R. (1992) 'A myth institutionalised: new social movements in Germany', *European Journal of Political Research* 21: 357–83.

Schnaiberg, A. (1980) *The Environment: From Surplus to Scarcity*, New York: Oxford University Press.

Schoon, N. (1998) 'The fight for green Britain', *Independent*, 17 January: 1.

Schumaker, P. (1975) 'Policy responsiveness to protest-group demands', *Journal of Politics* 37: 488–521.

Scott, A. (1990) *Ideology and the New Social Movements*, London: Unwin Hyman.
Sculthorpe, H. (1993) *Freedom to Roam*, London: Freedom Press.
Searle, D. (ed.) (1997) *Gathering Force*, London: The Big Issue.
Sharpe, T. (1975) *Blott on the Landscape*, London: Secker & Warburg.
Shaw, R. (1996) *The Activist's Handbook*, Berkley: University of California Press.
Shipley, P. (1976) *Revolutionaries in Modern Britain*, London: Bodley Head.
Short, J. (1991) *Imagined Country*, London: Routledge.
Sieghart, P. (ed.) (1979) *The Big Public Inquiry*, London: Council for Science and Society.
Smith, P. (1987) 'Thousands demand nuclear-free Britain', *Peace News* 2291: 3.
Smulian, M. (1994) 'DoT pays out to beat protesters', *Construction News* 6371: 1.
Snow, D. and Benford, R. (1988a) 'Frame alignment processes, micromobilization, and movement participation', *American Sociological Review* 51: 464–81.
—— (1988b) 'Ideology, frame resonance, and participant mobilization', in B. Klandermans, H. Kriesi and S. Tarrow (eds) *From Structure to Action: Comparing Social Movement Research Across Cultures*, International Social Movement Research, vol. 1.
—— (1992) 'Master frames and cycles of protest', in A. Morris and C. Mueller (eds) *Frontiers in New Social Movement Theory*, New Haven, CT: Yale University Press.
—— (1994) 'Identity fields: framing processes and the social construction of movement identities', in E. Larana, H. Johnston and J. Gusfield (eds) *New Social Movements: From Ideology to Identity*, Philadelphia: Temple University Press.
Solesbury, W. (1976) 'The environmental agenda; an illustration of how situations may become political issues and issues may demand responses from government: or how they may not', *Public Administration* 54: 379–97.
Soule, M. and Lease, G. (eds) (1995) *Reinventing Nature?*, Washington, DC: Island Press.
South, N. (1988) *Policing for Profit*, London: Sage.
Spretnak, C. and Capra, F. (1986) *Green Politics: The Global Promise*, London: Paladin.
Staggenborg, S. (1993) 'Critical events and the mobilization of the pro-choice movement', *Research in Political Sociology* 6: 319–45.
Starkie, D. (1982) *The Motorway Age*, London: Pergamon.
Steinmetz, G. (1994) 'Regulation theory, post-Marxism and the new social movements, *Comparative Studies in Society and History* 36(1): 176–212.
Stewart, J., Bray, J. and Must, E. (1995) *Road Block: How People Power Is Wrecking the Roads Programe*, London: Alarm UK.
Strang, D. and Meyer, J. (1993) 'Institutional conditions for diffusion', *Theory and Society* 22: 487–511.
Stryker, S. (1981) 'Symbolic interactionism: themes and variations', in M. Rosenberg and R. Turner (eds) *Social Psychology: Sociological Perspectives*, New York: Basic.
Styles, P. (1996) 'What's around the next bend?', *Red Pepper* 24: 14–16.
Sullivan, A. (1985) *Greening the Tories*, London: Centre for Policy Studies.
Swidler, A. (1986) 'Culture in action: symbols and strategies', *American Sociological Review* 51(2): 273–86.
—— (1994) 'Cultural power and social movements', in H. Johnston and B. Klandermans (eds) *Social Movements and Culture*, London: UCL Press.
Sylvan, D. (1985a) 'A critique of deep ecology: part one', *Radical Philosophy* 40: 2–12.
—— (1985b) 'A critique of deep ecology: part two', *Radical Philosophy* 41: 10–22.
Tarrow, S. (1994) *Power in Movement*, London: Cambridge University Press.
Taylor, B. (1991) 'The religion and politics of Earth First!', *The Ecologist* 21(6): 258–66.
—— (ed.) (1995) *Ecological Resistance Movements*, New York: State University of New York Press.

Taylor, D. (1992) 'Can the environmental movement attract and maintain the support of minorities?', in B. Bryant and P. Mohai (eds) (1992) *Race and the Incidence of Environmental Hazards*, Boulder, CO: Westview.

Taylor, R. (1985) 'Green politics and the peace movement', in D. Coates (ed.) *A Socialist Anatomy of Britain*, Cambridge: Polity Press.

Taylor, R. and Pritchard, C. (1980) *The Protest Makers*, London: Pergamon Press.

Taylor, R. and Young, N. (1987) 'Britain and the international peace movement in the 1980s', in R. Taylor and N. Young (eds) *Campaigns for Peace*, Manchester: Manchester University Press.

Touraine, A. (1981) *The Voice and the Eye*, Cambridge: Cambridge University Press.

—— (1983) *Anti-nuclear Protest*, Cambridge: Cambridge University Press.

Thomas, L. (1971) 'Family correlates of student political activism', *Developmental Psychology* 4: 206–14.

Thornton, S. (1994) 'Moral panic, the media and British rave culture', in A. Ross and T. Rose (eds) *Microphone Fiends*, London: Routledge.

Tickell, O. (1993) 'Direct action is the only way left', *Guardian*, 28 May, section 2: 2.

Tilly, C. (1978) *From Mobilization to Revolution*, Reading, MA: Addison-Wesley.

—— (1986) *The Contentious French*, Cambridge, MA: Harvard University Press.

Toker, B. (1997) *Earth for Sale*, Boston, MA: South End Press.

Torrance, J. (1991) 'Carmageddon', *Green Line* 91: 13.

Trainer, F.E. (1985) *Abandon Affluence!*, London: Zed.

Travis, A. (1994) 'Eco-warriors to step up battle against Howard's bill', *Guardian*, 6 August: 5.

Twinn, I. (1978) *Public Involvement or Public Protest: A Case Study of the M3 at Winchester 1971–1974*, London: Polytechnic of the South Bank.

Tyme, J. (1978) *Motorways versus Democracy*, London: Macmillan.

van der Heijdan, H.-A., Koopmans, R. and Giugni, M. (1992) 'The West European environmental movement', in M. Finger (ed.) *The Green Movement Worldwide*, Greenwich, CT: JAI Press.

Veldman, M. (1994) *Fantasy, the Bomb and the Greening of Britain*, London: Cambridge University Press.

Vidal, J. (1992) 'Last ditch stand on Cobbett's patch', *Guardian*, 20 March, section 2: 1.

—— (1993) 'Explode a condom save the world', *Guardian*, 10 July, section 2: 2.

—— (1997) *McLibel*, London: Macmillan.

Vincent, A. (1993) 'The character of ecology', *Environmental Politics* 2(2): 248–76.

Volmer, C. (1997) *Unlocking the Gridlock*, London: Friends of the Earth.

Wall, D. (1989) 'The Greenshirt effect', *Searchlight* June: 4–5.

—— (1990) *Getting There*, London: Greenprint.

—— (1991a) 'Goodbye to the Green Party', *Green Line* 88: 14–15.

—— (1991b) 'Environmental', *Venue* 246: 10–13.

—— (1994a) *Green History*, London: Routledge.

—— (1994b) *A Bower Against Endless Night – A History of the UK Green Party*, London: Green Party.

Walker, A. (1988) *Living by the Word*, London: Women's Press.

Walsh, E. (1988) *Democracy in the Shadows: Citizen Mobilization in the Wake of the Accident at Three Mile Island*, Westport, CT: Greenwood Press.

Walsh, E. and Warland, R. (1983) 'Social movement involvement in the wake of a nuclear accident: activists and free riders in the Three Mile Island Area', *American Sociological Review* 48: 764–81.

Ward, H. (1983) 'The anti-nuclear movement', in D. Marsh (ed.) *Pressure Politics*, London: Junction.
Ward, H. and Samways, D. (1992) 'Environmental policy', in D. Marsh and R. Rhodes (eds) *Implementing Thatcherite Policies*, Milton Keynes: Open University Press.
Watkins, A. (1991) *A Conservative Coup*, London: Duckworth.
Weale, A. (1992) *The New Politics of Air Pollution*, Manchester: Manchester University Press.
Weiss, R. (1994) *Learning from Strangers*, New York: Free Press.
Welsh, I. (1993) 'The NIMBY syndrome', *British Journal of the Society for the History of Science* 26(1): 15–33.
—— (1994) 'Risk, reflexivity and the globalisation of environmental politics', Social, Economic and Political Environment Group, University of the West of England, Bristol.
Welsh, I. and McLeish, P. (1996) 'The European road to nowhere', *Anarchist Studies* 4: 27–44.
Weston, J. (1989) *The FoE Experience*, Oxford: School of Planning, Oxford Polytechnic.
Whitelegg, J. (1989) 'Transport policy: off the rails', in J. Mohan (ed.) *The Political Geography of Contemporary Britain*, London: Macmillan.
Wiesenthal, H. (1993) *Realism in Green Politics*, Manchester: Manchester University Press.
Wilkinson, P. and Schofield, J. (1994) *Warrior*, Cambridge: Lutterworth Press.
Williams, H. (1991) *Autogeddon*, London: Jonathon Cape.
Wilson, D. (1984) 'FoE: today and tomorrow', in D. Wilson (ed.) *The Environmental Crisis*, London: Heinemann.
Wiltfang, G. and McAdam, D. (1991) 'Distinguishing cost and risk in sanctuary activism', *Social Forces* 69: 987–1010.
Winch, P. (1958) *The Idea of a Social Science and its Relation to Philosophy*, London: Routledge & Kegan Paul.
Wistrich, E. (1983) *The Politics of Transport*, London: Longman.
Witherspoon, S. and Martin, J. (1992) 'What do you mean by green?', in R. Jowell, L. Brook, G. Prior and B. Taylor (eds) *British Social Attitudes: The 9 Report*, Aldershot: Dartmouth.
Wolmer, C. (1997) *Unlocking the Gridlock*, London: Friends of the Earth.
Working Group on Pop Festivals (1976) *Free Festivals*, London: HMSO.
Worster, D. (1977) *Nature's Economy*, London: Cambridge University Press.
Young, S. (1992) 'The different dimensions of green politics', *Environmental Politics* 1(1): 9–44.
Zakin, S. (1993) *Coyotes and Town Dogs*, New York: Penguin.
Zald, M. (1992) 'Looking backward to look forward: reflections on the past and future of the resource mobilization research program', in A. Morris and C. Mueller (eds) (1992) *Frontiers in New Social Movement Theory*, New Haven, CT: Yale University Press.
—— (1996) 'Culture, ideology and strategic framing', in D. McAdam, J. McCarthy and M. Zald (eds) *Comparative Perspectives on Social Movements*, Cambridge: Cambridge University Press.
Zald, M. and McCarthy, J. (1979) *The Dynamics of Social Movements*, Cambridge, MA: Winthrop.
—— (1987) *Social Movements in an Organizational Society*, New Brunswick, NJ: Transaction Books.
Zimmerman, M. (1994) *Contesting the Earth's Future*, Berkley: University of California Press.

Zisk, B. (1992) *The Politics of Transformation – Local Activism in the Peace and Environmental Movements*, Westport, CT: Praeger.
Zukier, H. (1982) 'Situational determinants of behaviour', *Social Research* 49: 1073–91.

Index

For Product Safety Concerns and Information please contact our EU
representative GPSR@taylorandfrancis.com
Taylor & Francis Verlag GmbH, Kaufingerstraße 24, 80331 München, Germany

www.ingramcontent.com/pod-product-compliance
Ingram Content Group UK Ltd.
Pitfield, Milton Keynes, MK11 3LW, UK
UKHW021721240926
475702UK00029BA/1744
* 9 7 8 0 4 1 5 8 6 2 7 0 7 *